MANUAL OF GREY WATER TREATMENT PRACTICE

Part I

On-Site Treatment
and Subsurface Disposal

Part II

Characterization of
Grey Water and
Soil Mantle Purification

edited by
John H. Timothy Winneberger

ANN ARBOR SCIENCE
P.O. BOX 1425 • ANN ARBOR, MICHIGAN 48106

Published by
ANN ARBOR SCIENCE PUBLISHERS, INC.
P.O. Box 1425, Ann Arbor, Michigan 48106

Library of Congress Catalog Card Number 76-5236
ISBN 0-250-40136-3

Printed in the United States

PREFACE TO PART I

For over one hundred years since its initial widespread use in London, the water flushing toilet always has used substantial amounts of pure drinkable water to transport relatively minuscule quantities of body waste through a network of sewer pipes to rivers, lakes or oceans (with and without intervening treatment) or to an individual, on-site treatment and/or disposal system. Mixed with the body waste is wastewater from baths, sinks and other water-using household appliances and fixtures as well as industrial waste in the case of centralized sewerage systems.

Recent advances in sanitation engineering technology have raised serious questions concerning such methods. For example, it is reasonable to ask why 40% of our invaluable drinking water should be wasted as a carrier for domestic body waste. One also can question why body waste, containing as it does pathogenic disease-causing bacteria and viruses, should be co-mingled with so-called "grey water" from baths, sinks and appliances. Where treatment plants exist, great pains are exerted to recover the useful water in a form suitable for either reuse or harmless discharge into the environment. Usually these efforts are less than complete, and our rivers, streams, lakes and ground water tables are encountering substantial pollution. In most parts of the world, this entire mixture still is dumped without treatment directly into the environment.

Among the first to recognize the advantage of separating grey water from toilet "black waste" were the French, who have pioneered many advances in the arts of sanitary engineering and plumbing. Particularly in areas remote from centralized sewerage networks, the French mandate* the use of separate collection systems for both black waste and grey water with different treatments stipulated for each. With the recent advent of several types of flushing toilets that employ a reusable non-aqueous fluid for transporting waste to a collection point, it is now possible in many locales to eliminate entirely the previously universal use of drinking water for toilet flushing and to deal with the problems of human waste ("black waste") and grey water treatment and disposal separately.

Key to the new waterless flushing toilet concept is the collection of the relatively small amount of body waste in a compact sealed tank where it is held quiescently for long periods of time until it can be drawn off by a pump truck

---

*REGULATIONS & INFORMATION CONCERNING SANITATION FOR SINGLE FAMILY HOMES & SMALL COMMUNITIES, National Association of Public Health Enterprises and Industries, Paris 8^{e}, 3d. Ed. – 1969

and transported to a central station for either reclamation of valuable byproducts or ultimate disposal. When utilized on a plot remote from centralized sewerage, such a system eliminates the need for on-site black waste treatment and disposal and leaves only the problem of handling the remaining grey water in a manner conforming to good sanitation and health practices.

Being so different in composition as compared to combined waste from conventional systems, grey water entirely free of black waste pollutants is more acceptable esthetically for on-site treatment and far more economical to process in terms of equipment and space required. For one thing, the quantity of fluid to be treated in an average home equipped with a waterless flushing toilet is reduced over conventional systems by some 40% and even more in installations such as public restrooms, where the only grey water often is from a simple sink fixture. For another, the containment of black waste means that all of its disease-causing micro-organisms and polluting nitrates do not enter the grey water system or the subsurface disposal field. Thus it can be seen that the amalgamation of waterless flushing toilet techniques with separate systems for treating black waste and grey water present potential users, cognizant governmental authorities, and legislative bodies with a viable alternative to conventional wastewater handling and treatment practices. Whereas the present choice is between centralized sewage systems, which never restore the water to its original condition, and septic tanks, some 15 million of which are now employed in the United States alone with questionable safety and reliability, the possibility of providing modern safe sanitation in new towns and cities with these advanced techniques is worthy of the most careful study.

This manual has been prepared as an aid to those who wish to employ the new "sewerless" concepts. It deals exclusively with techniques for treating grey water and then discharging it on-site in a manner that will insure the health and safety of the people and communities involved. Edited by Dr. John T. Winneberger, a recognized authority in the field of subsurface wastewater disposal systems and former member of the academic research staff of the University of California at Berkeley, the MANUAL OF SEPTIC-TANK PRACTICE, a classic treatise by the United States Public Health Service, has been scaled down to accommodate the lesser problem of management of grey water.

Being so widely used, the MANUAL OF SEPTIC-TANK PRACTICE is well-known to the myriads of people responsible for on-site treatment and disposal of wastewater. Fortunately, virtually all of the information and technology employed in the selection, installation and operation of septic tanks and disposal fields is directly adaptable as is to the treatment and disposal of grey water, with the major exception being that smaller treatment devices and disposal fields can now be employed. The size reduction of a subsurface disposal field now possible of achievement is of major economic import to both land owners and regulatory authorities in communities heretofore requiring plots of a certain size as a minimum. The requirement for periodic service of the waterless flushing toilet and its waste collection system overcomes one of the major disadvantages of present septic tanks, namely they require periodic inspection and maintenance but seldom receive it. Now the grey water systems can be maintained at the same time.

While it must be emphasized both that this present treatise does not carry the endorsement of the United States Public Health Service, nor that of any local administrative authorities responsible for approving on-site waste disposal systems, its recommended practices conform closely to those that have stood the test of time and are already widely accepted. The major changes involving reduced sizing of various components evolved from these recent advances in the state-of-the-art.

We wish to acknowledge the contributions of the United States Public Health Service and its Joint Committee on Rural Sanitation to the preparation of the authoritative MANUAL OF SEPTIC-TANK PRACTICE and to M.L. Langlade, Director of the French National Association of Public Health, who edited the authoritative treatise on the same subject used in his country.

Robert L. Rod
Corporate Science Director
Monogram Industries, Inc.

Santa Monica, California

PREFACE TO PART II

Beginning several years ago when the new subject of the "sewerless society" was first discussed as a viable alternative to centralized sewers, proponents of this innovative approach of necessity have expended substantial time and effort properly briefing cognizant health and public safety officials on various aspects of hardware as well as answering scientific questions of concern. Primary interest increasingly has been concerned with newly introduced systems separating toilet waste from other wastewater and then handling and treating the two separately. The terms "black wastes" and "grey water" have become widely understood as has the advantage of separating the wastewaters they describe at the onset.

As with any new technology, the body of knowledge grows exponentially. From what were as best "black arts," the twin subjects of on-site sewage treatment and the sewerless society have become recognized sciences in a space of only five years. No longer can an underutilized grave digger "moonlight" with impunity excavating septic tank pits and leach fields, simply because the authorities now understand that these are potential health hazards and nuisances in many parts of the country and abroad. Regulations as to their sizing and construction increasingly are being implemented and enforced in the hope that their associated hazards can be minimized, but time has shown that such systems can and never will prove foolproof in their present configurations.

The main health and reliability problems associated with septic tank systems (and their disposal fields) are caused by introduction of human waste to devices and soil systems that cannot always be guaranteed to remove pathogenic organisms or prevent soil clogging in the first place. On the other hand, it is now increasingly clear that the efficiency and life of a subsurface soil disposal system can be enhanced and its size reduced by eliminating from the system the human body waste products and/or by reducing the flow. The simplest analogy is first passing a pint of tomato juice through a kitchen strainer and then comparing results when one tries over again with two pints of a thick, gooey tomato paste.

Once this is understood, it becomes possible to better utilize the soil system for on-site waste disposal and thereby obviate the need for complex community sewerage tying together all the homes and industrial facilities with one expensive and often inefficient centralized wastewater treatment plant. Properly treated wastewater belongs in the ground. There is no earthly reason to transport it great distances to some central point where the vast quantity collected complicates its natural return to the environment by overloading and polluting the receiving body.

The alternative that holds promise of transforming society from one of a few huge megalopolises, each concentrated around a central sewage treatment plant, to one better utilizing our vast amounts of unused but otherwise productive land is properly accomplished on-site subsurface disposal of wastewaters. Because present subsurface systems involving commingled toilet and other wastes more often than not are incapable of proper hygiene and operation, it goes without saying that systems separating the two should be preferred in the United States as they have been in France for many years.

Much information is available elsewhere about waterless flushing toilets. These will once and for all remove from consideration questions of on-site disposal of body waste and leave only problems relating to disposal of the non-toilet soils we call "grey water." The strength of grey water and its volume can vary widely. In a public restroom at a remote park, grey water usually is nothing more than harmless handwashing wastewater with mixed-in soap. In a home or restaurant replete with garbage disposal, grey water has a far greater volume per capita and, clearly, substantially more strength.

The question then arises as to what strength and volume of grey water one can dispose of on-site? The first of several partial answers to the question must be the characterization of the particular grey water. The second is its expected volume over a typical period of use, including peaks, valleys and surges. The third answer will depend upon the character of the soil into which the wastewater will be discharged. And the fourth, and by no means the least important, will be the ground discharge standards prevailing in a locale.

But where are the data needed to answer the questions? As anyone knowledgeable in the field will agree, data on characterization of black waste and grey water have been notable by their absence. Information on flow traditionally has been on the basis of bedrooms, not people, and thus is unreliable at best. And the relative receptivity of soil to all the complex constituents of sewage has been condensed into a "go-no go" test based upon the rate at which fresh water percolates into a hole in the ground.

Facing this lack of facts, it was decided to rectify matters and put together in one volume up-to-date information from a well-known group of authorities in the field of sewerless sanitation and subsurface soil disposal systems. As the reader will find, there are answered herein in concise form questions concerning the expected strength and flow of wastewaters from various types of homes and household fixtures. Also included is a treatise on what the soil system is capable of accomplishing by way of waste purification once its mechanisms are understood. And the advantages of discharging only grey water into the ground are covered in detail. All of this information we hope will prove invaluable to sanitary and environmental engineers, sanitarians and health officials. In the course of their duties, these professionals will encounter situations where the treatment and disposal of grey water may be as simple as spraying it on a lawn as it arrives from the fixtures, or evaporating it from solar ponds.

At the opposite extreme will be sophisticated on-site grey water treatment devices which by law may have to produce an effluent of drinking water quality. In between will be the majority of systems making use of minimal settling pretreatment entirely suitable for use with reduced size subsurface soil fields having vastly extended life and efficiency. Part 1 of this series of manuals discussed such on-site systems and their scaling.

This booklet comprises Part 2 and deals heavily with the characterization of grey water as developed by Dr. Rein Laak, Associate Professor of Civil Engineering at the University of Connecticut, and by Warren D. Hypes and his associates, of the National Aeronautics and Space Administration's Langley Research Center, in Hampton, Virginia. Both authors discuss typical test homes monitored carefully for their grey water characterizations, with Professor Laak giving authoritative data not only on chemical composition but flow from typical household fixtures including toilets. Mr. Hypes presents in turn the most comprehensive chemical and biological analyses of grey water with and without garbage grindings ever undertaken anywhere up to this time. The NASA data were obtained with the same types of sophisticated analytical equipment that served the APOLLO man-on-the-moon program. These results are a particularly worthwhile transfer of space technology into more mundane but highly important applications here on earth.

Rounding out this volume is an excellent paper on the physical and chemical properties of soil systems showing that nature is fully capable of doing its job if mankind will only work within the "rules." Its author is Dr. Kenneth Y. Chen, Associate Professor of Environmental Engineering Programs of the University of Southern California's School of Engineering in Los Angeles. Most soil systems and the friendly bacteria residing therein can treat wastewater and return the pollutants they contain back to the ecocycle as useful and harmless matter when given a chance. It thus behooves the designer of a proper sewerless home or factory or roadside station to look first to the left and estimate the grey water quality and quantity he or she will encounter and then to the right to determine the absorptive character of the particular soil. Then the design of the intermediate treatment system, if any, should be as safe, inexpensive and reliable as our now expanded state-of-the-art permits.

All of this useful material has been collected, edited and commented upon by one of the outstanding authorities on the sewerless society and subsurface waste disposal soil systems, Dr. John "Tim" Winneberger, of Berkeley, California, who performed a similar contribution in Part 1 of the series. Readers will be treated in this booklet's opening section to some of Tim's classic comments on our society. These invariably have brightened otherwise deadly serious technical meetings on subsurface waste disposal in which he participates. Winneberger's Law II given herein is priceless; history records his first law in Vol. 34, No. 1, of the *Journal of Environmental Health.*

The authors' opinions are theirs alone and neither represent those of their institutions or of Monogram Industries, Inc. Thanks are due to each of the authors and their associates and to the Universities of Connecticut and Southern California and the National Aeronautics and Space Administration for permitting publication of the accompanying articles.

ROBERT L. ROD
Vice President - Science
Monogram Industries, Inc.

Santa Monica, California

TABLE OF CONTENTS

PART I On-Site Treatment and Subsurface Disposal

Page

APPENDICES

PART II Characterization of Grey Water and Soil Mantle Purification

Page

Part I

On-Site Treatment and Subsurface Disposal

ON-SITE TREATMENT AND SUBSURFACE DISPOSAL

J.T. Winneberger

INTRODUCTION

The essence of pollution is concentration of almost anything to the point that it alters a local environment in a manner mankind deems undesirable. It has been said that the solution to pollution is dilution, and, indeed, it is. In recent years, however, mankind's ability and inclination to concentrate wastes have overloaded many local environments where adequate dilution is no longer possible.

One solution to pollution is on-site waste management as opposed to collection of small volumes into great volumes. This approach has heretofore been technically underdeveloped and unsophisticated. Cesspools and septic tanks are still pretty much a backyard affair. Nevertheless, authorities are beginning to realize that the soil mantle has much to offer as a treatment system and the future is beginning to see development of professional design and management of such devices.

A solution to water pollution is refraining from discharging wastes into clean waters from the onset. Collection of black wastes, undiluted with water heretofore required for flushing toilets, is a major step in that direction. This results in lower wastewater volumes of less noxious wastes being produced from a household. Management of these volumes then becomes a reduced problem.

Inasmuch as scientists do not yet fully understand subsurface waste-water disposal systems, they diverge in their recommendations for practices. On top of this, the practical world offers its own limitations. Thus, it is contemplated to discuss grey water management in more than one part. This Part I is limited to the presentation of wastewater treatment and subsurface wastewater disposal practices as described by the United States Public Health Service. The MANUAL OF SEPTIC-TANK PRACTICE, upon which Part I is based, has been scaled down to accommodate management of grey water. Thus, the reader can use this volume in designing a grey water system in reasonable conformance with national standards.

Part II of this manual is a recent compilation of data characterizing grey water from various types of plumbing fixtures and water-using appliances in typical residential and public sanitation systems. For the first time, availability of detailed biological and chemical analyses of the wastewaters from each such fixture and appliance used with typical household soaps, detergents, toothpastes, cooking oils, and the like, provide meaningful "input"

John Timothy Winneberger, Ph.D., Consultant, Individual Sanitation Systems, 1018 Hearst Avenue, Berkeley, California 94710.

data needed to design a treatment system capable of providing an effluent meeting prescribed standards.

Commonly, subsurface disposal systems serve in non-sewered areas, and a better understanding of the soil mantle purification system is a valuable aid in designing the overall grey water treatment system. Thus Part II also includes extracts of various technical reports from competent governmental research agencies, which prove conclusively that a properly designed and installed subsurface disposal system eliminates ground pollution. In particular, where the soil is suitably pervious, normal bacterial activities are capable of reducing biochemical oxygen demand to the tertiary level, while at the same time, nutrients found in detergents are degraded almost 100%. In non-water carrier-type toilet systems, wherein body wastes containing polluting nitrates are stored for subsequent off-site disposal elsewhere, the problem of treating grey water containing little if any nitrates is considerably simplified. All of these factors have been taken into consideration during the preparation of Part I of this MANUAL OF GREY WATER TREATMENT PRACTICE to the point where one can be assured that the recommendations contained herein are valid for treating grey water where soil conditions meet the requirements.

It is necessary here to point out that presentation of Part I is not intended by the editor or Monogram Industries, Inc. to carry the endorsement of treatment and subsurface wastewater disposal practices of the United States Public Health Service. Also, the reader should first accommodate the requirements of his local authorities.

DEFINITIONS

Black Waste—Liquid and solid body waste from a toilet (including extraneous debris).

Black Water—Body waste and water used for flushing and/or transport.

Building Drain—That part of the lowest piping of a drainage system which receives waste discharges from other drainage pipes of the building and conveys them to the building sewer which begins three feet outside of the building wall.

Building Sewer—That part of a drainage system which extends from the end of the building drain and conveys its discharges to a public sewer, private sewer, individual sewage disposal system, or other point of disposal.

Disposal Bed—A flat excavation over 36" in minimum dimension, with a minimum of 12" of clean aggregate containing a system of distribution pipes, and covered with a minimum of 12" of earth cover.

Disposal Field—A system of disposal trenches, disposal beds, or disposal pits, or combinations of these devices, designed for the subsurface disposal of treated grey waters into soils. (Disposal fields and their devices are commonly termed; seepage fields, leach fields, tile fields, drain fields, septic fields, absorption fields, sometimes the word "cesspool" is misapplied, and others.)

Disposal Pit—A comparatively deep disposal trench, or disposal bed, or a cylindrical excavation, either backfilled with aggregate, or provided with specially fabricated liners.

Disposal Trench—A trench not over 36" in width, with a minimum of 12" of clean aggregate containing a distribution pipe, and covered with a minimum of 12" of earth cover.

Distribution System—A system of perforated distribution pipes or drain tiles intended to distribute treated grey waters throughout a disposal field.

Drainage Fixture Unit Value—A common measure of the probable discharge into a drainage system from various types of plumbing fixtures. This value for a particular fixture depends on its volume rate of drainage discharge, on the time duration of a single drainage operation, and on the average time between successive operations.

Grey Water—Liquid and solid wastes from fixtures and water-using appliances other than black wastes as defined above.

Grey Water Treatment Tank—A water-tight, covered receptacle designed and constructed to receive the discharge of grey waters from a building sewer, separate solids and grease from the liquids, store or digest separated solids and grease, optionally eliminate any other undesired pollutants, and allow the clarified liquids to discharge to final disposal.

Individual Sewage Disposal System—A privately owned and maintained system of sewage treatment and disposal facilities serving a single lot.

Parallel Distribution—Components of a disposal field arranged such that each is served liquid in equal quantities.

Scum-Wastes—Those which float on the surface of grey waters.

Scum Clear Space—Distance between the bottom of the scum mat and the bottom of the outlet device of a grey water treatment tank.

Serial Distribution—Components of a disposal field arranged such that each is forced to pond in turn utilizing the total effective absorption area, before liquid flows into the succeeding component.

Sewage—Any liquid waste containing animal or vegetable matter in suspension or solution, and may include liquids containing chemicals in solution.

Sludge—The accumulated settled solids deposited from sewage.

Sludge Clear Space—The distance between the top of the sludge and the bottom of the outlet device of a grey water treatment tank.

Standard Disposal Trench—A trench 12" to 36" in width containing 12" of clean, coarse aggregate and a distribution pipe, covered with a minimum of 12" of earth cover.

SUITABILITY OF SOIL

The first step in the design of a disposal field is to determine whether or not the soil is suitable for the absorption of treated grey waters and, if so, how much area is required. The soil must have an acceptable percolation rate; also ground waters and impervious strata must not interfere with functions of the disposal field. In general, two conditions must be met:

(1) The percolation time should be within the range of those specified in Table 1, page 8.

(2) The bottommost surfaces of a disposal field should be at least 4' above the maximum seasonal elevation of the ground water table, rock formations, or other impervious strata.

Unless these conditions can be satisfied, the site is unsuitable for a conventional subsurface grey water disposal system.

PERCOLATION TESTS

Subsurface explorations are necessary to determine subsurface formations in a given area. An auger with an extension handle, as shown in Figure 1 (p. 5), is often used for making the investigation. In some cases, an examination of road cuts, stream embankments, or building excavations will give useful information. Wells and well drillers' logs can also be used to obtain information on ground water and subsurface conditions. In some areas, subsoil strata vary widely in short distances, and borings must be made at the site of the system. If the subsoil appears suitable, as judged by other characteristics described in Appendix A, percolation tests should be made at points and elevations selected as typical of the area in which the disposal field will be located.

The percolation tests help to determine the acceptability of the site and establish the design size of the subsurface disposal field. The length of time required for percolation tests will vary in different types of soil. The safest method is to make tests in holes which have been kept filled with water for at least 4 hours, preferably overnight. This is particularly desirable if the tests are to be made by an inexperienced person, and in some soils it is necessary even if the individual has had considerable experience (as in soils which swell upon wetting). Percolation rates should be figured on the basis of the test data obtained after the soil has had opportunity to become wetted or saturated and has had opportunity to swell for at least 24 hours. Enough tests should be made in separate holes to assure that the results are valid.

The percolation test developed at the Robert A. Taft Sanitary Engineering Center incorporates these principles. Its use is particularly recommended when knowledge of soil types and soil structure is limited. When previous experience and information on soil characteristics are available, some persons prefer other percolation test procedures, such as those developed by Kiker and by Ludwig which are cited in Appendix A.

Procedure for Percolation Tests Developed at Robert A. Taft Sanitary Engineering Center

1. *Number and location of tests.*—Six or more tests shall be made in separate test holes spaced uniformly over the proposed disposal field site.

2. *Type of test hole.*—Dig or bore a hole, with horizontal dimensions of from 4 to 12 inches and vertical sides to the depth of the proposed disposal field. In order to save time, labor, and volume of water required per test, the holes can be bored with a 4 inch auger. (See Fig. 1, page 5.)

3. *Preparation of test hole.*—Carefully scratch the bottom and sides of the

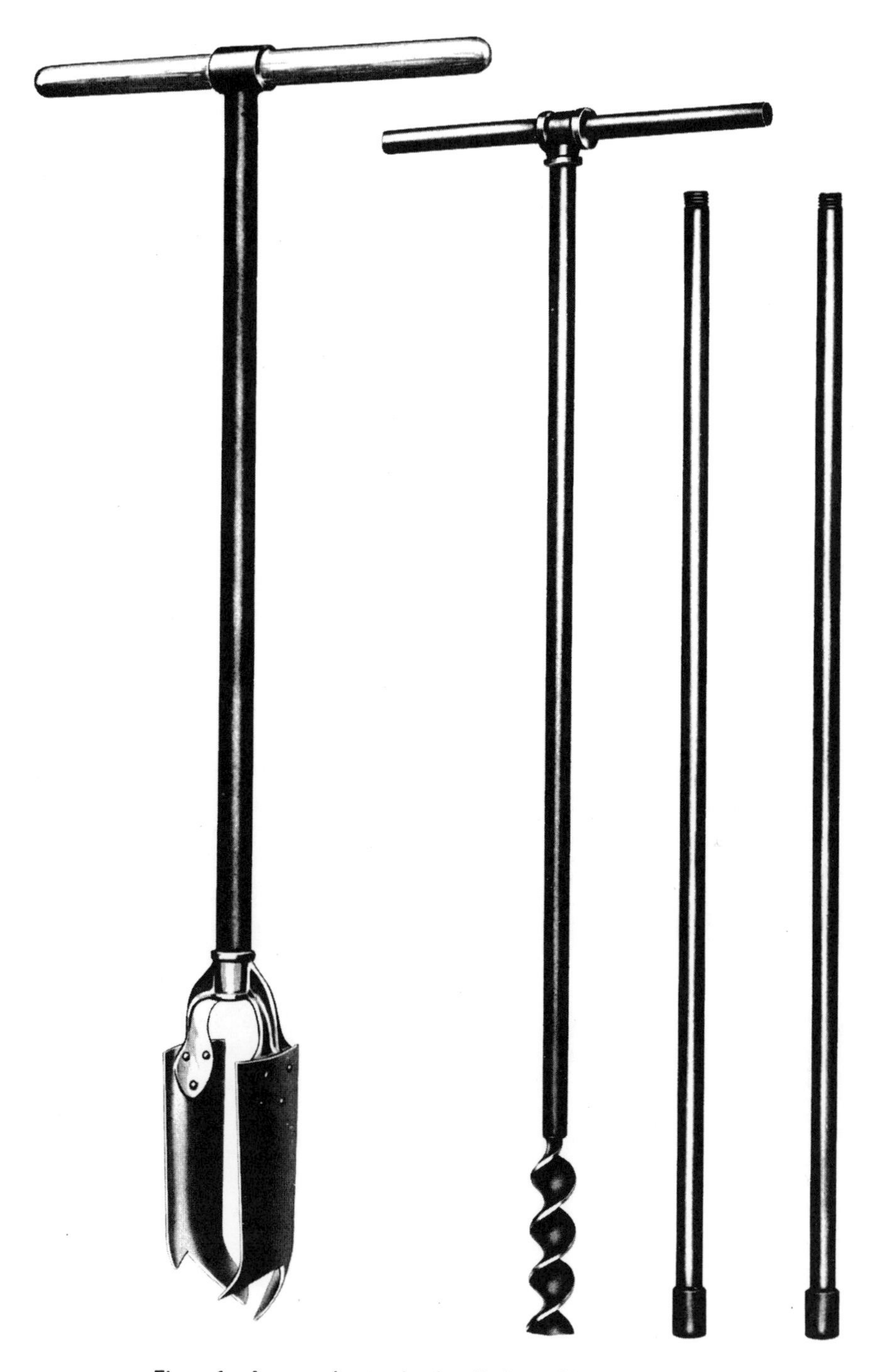

Figure 1.—Auger and extension handle for making test borings.

hole with a knife blade or sharp-pointed instrument, in order to remove any smeared soil surfaces and to provide a natural soil interface into which water may infiltrate. Remove all loose material from the hole. Add 2 inches of coarse sand or fine gravel to protect the bottom from scouring and sediment.

4. *Saturation and swelling of the soil.*–It is important to distinguish between saturation and swelling. Saturation means that the void spaces between soil particles are full of water. This can be accomplished in a short period of time. Swelling is caused by intrusion of water into the individual soil particle. This is a slow process, especially in clay-type soil, and is the reason for requiring a prolonged soaking period.

In the conduct of the test, carefully fill the hole with clear water to a minimum depth of 12 inches over the gravel. In most soils, it is necessary to refill the hole by supplying a surplus reservoir of water, possibly by means of an automatic syphon, to keep water in the hole for at least 4 hours and preferably overnight. Determine the percolation rate 24 hours after water is first added to the hole. This procedure is to insure that the soil is given ample opportunity to swell and to approach the condition it will be in during the wettest season of the year. Thus, the test will give comparable results in the same soil, whether made in a dry or in a wet season. In sandy soils containing little or no clay, the swelling procedure is not essential, and the test may be made as described under item 5C, after the water from one filling of the hole has completely seeped away.

5. *Percolation-rate measurement.*–With the exception of sandy soils, percolation-rate measurements shall be made on the day following the procedure described under item 4, above.

A. If water remains in the test hole after the overnight swelling period, adjust the depth to approximately 6 inches over the gravel. From a fixed reference point, measure the drop in water level over a 30 minute period. This drop is used to calculate the percolation rate.

B. If no water remains in the hole after the overnight swelling period, add clear water to bring the depth of water in the hole to approximately 6 inches over the gravel. From a fixed reference point, measure the drop in water level at approximately 30 minute intervals for 4 hours, refilling 6 inches over the gravel as necessary. The drop that occurs during the final 30 minute period is used to calculate the percolation rate. The drops during prior periods provide information for possible modification of the procedure to suit local circumstances.

C. In sandy soils (or other soils in which the first 6 inches of water seeps away in less than 30 minutes, after the overnight swelling period), the time interval between measurements shall be taken as 10 minutes and the test run for one hour. The drop that occurs during the final 10 minutes is used to calculate the percolation rate.

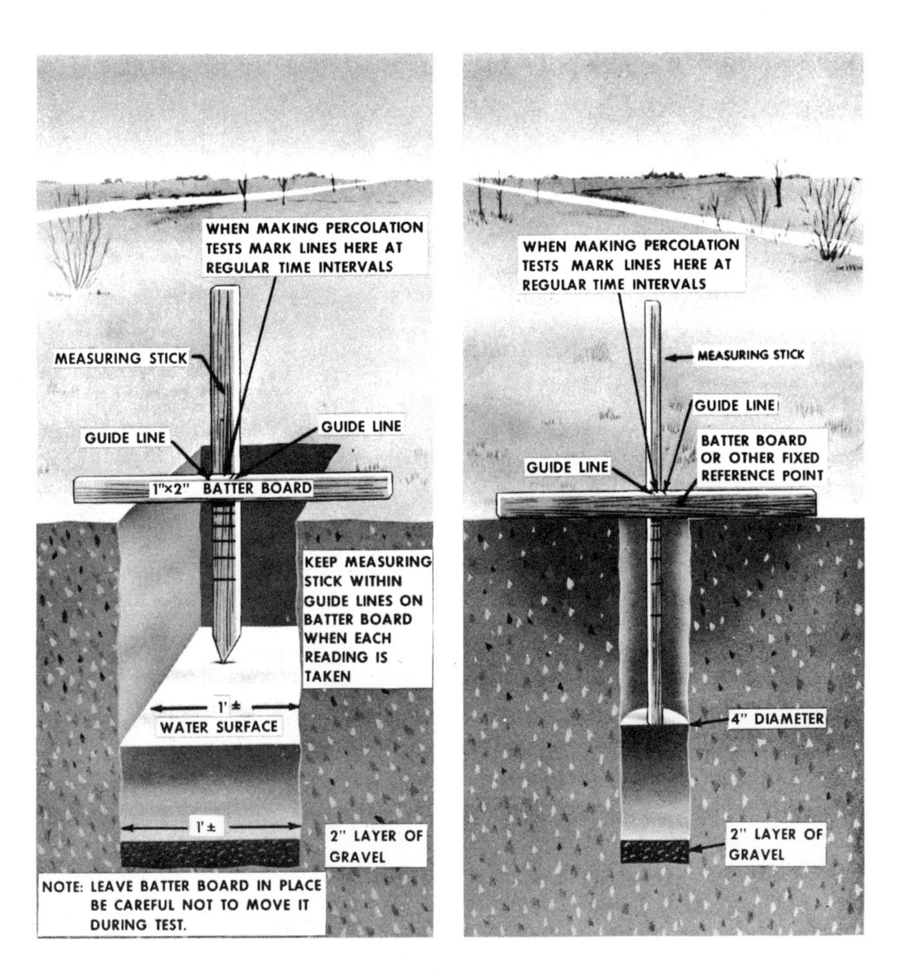

Figure 2.—Methods of making percolation tests.

Table 1.—Absorption-area requirements for individual residences (a)

Percolation rate (time required for water to fall one inch, in minutes)	Required absorption area, in sq. ft. per bedroom (b), standard trench (c), disposal beds (c), and disposal pits (d)	Percolation rate (time required for water to fall one inch, in minutes)	Required absorption area, in sq. ft. per bedroom (b), standard trench (c), disposal beds (c), and disposal pits (d)
1 or less	42	10	100
2	50	15	115
3	60	30 (e)	150
4	70	45 (e)	180
5	75	60 (e), (f)	200

(a) It is desirable to provide sufficient land area for entire new disposal field if needed in future.

(b) In every case sufficient land area should be provided for the number of bedrooms (minimum of 2) that can be reasonably anticipated, including the unfinished space available for conversion as additional bedrooms.

(c) Absorption area is figured as bottom areas of trenches or beds.

(d) Absorption area for pits is figured as effective side wall area beneath the inlet.

(e) Unsuitable for disposal pits if over thirty.

(f) Unsuitable for disposal trenches and beds if over sixty.

Note: Provides for garbage grinder and automatic washing machine. Maximum occupancy of about 3 persons per bedroom is assumed.

DISPOSAL FIELD SYSTEM

For areas where the percolation rates and soil characteristics are good, the next step after making the percolation tests is to determine the required absorption area for a private residence from Table 1 or Figure 3 (page 9), and to select the disposal field system that will be satisfactory for the area in question. As noted in Table 1, soil in which the percolation rate is slower than 1 inch in 30 minutes is unsuitable for disposal pits, and that slower than 1 inch in 60 minutes is unsuitable for other conventional disposal fields.

When considering public sanitation systems where grey water usage is usually far less than that encountered in a private residence, the absorption area will decrease directly in proportion to the reduced flow.

When a disposal field system is determined to be feasible, three types of design may be considered: disposal trenches, disposal beds, and disposal pits. A modification of the standard disposal trench is discussed on page 18 (Table 3) giving credit for more than the standard 12 inches of gravel depth in the trench.

The selection of the disposal field system will be dependent to some extent on the location of the system in the area under consideration. A safe distance should be maintained between the site and any source of water supply. Since the distance that pollution will travel underground depends upon numerous factors, including the characteristics of the subsoil formations and the quantity

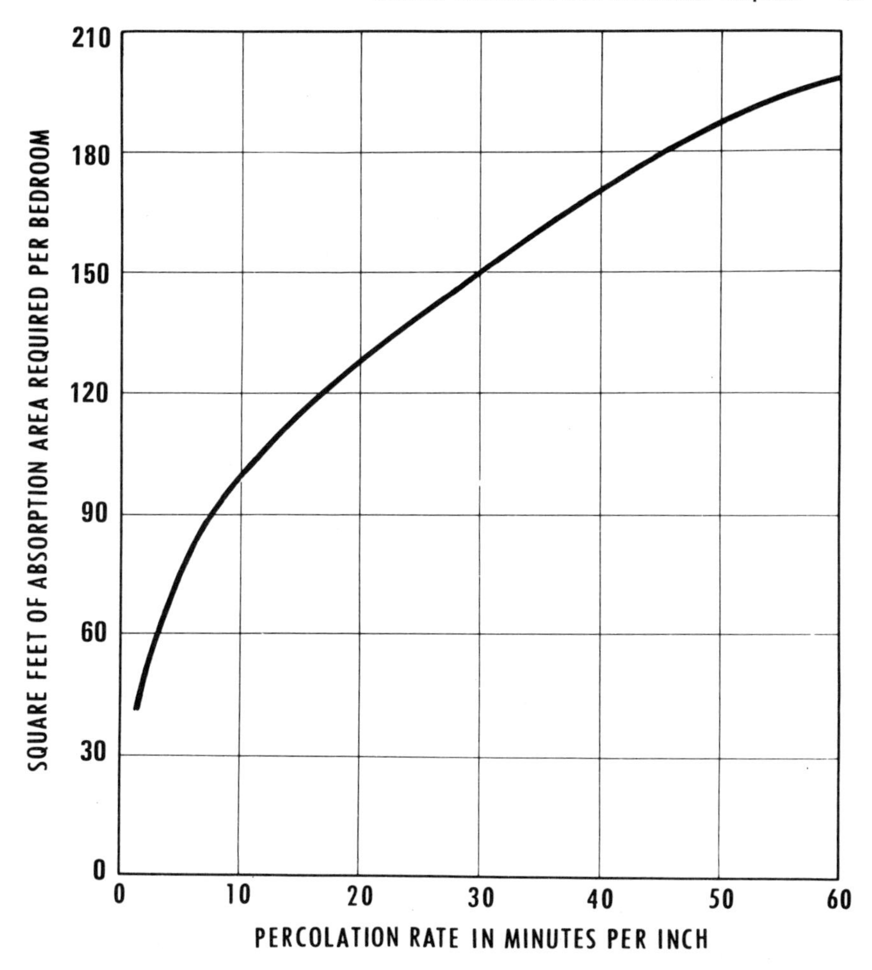

Figure 3.—Absorption area requirements for private residences.

of grey waters discharged, no specified distance would be absolutely safe in all localities. Ordinarily, of course, the greater the distance, the greater will be the safety provided. In general, location of components of grey water treatment disposal systems should be as shown in Table 2.

Disposal pits should not be used in areas where domestic water supplies are obtained from shallow wells, or where there are limestone formations and sinkholes with connection to underground channels through which pollution may travel to water sources.

Details pertaining to local water wells, such as depth, type of construction, vertical zone of influence, etc., together with data on the geological formations and porosity of subsoil strata, should be considered in determining the safe allowable distance between wells and subsurface disposal systems.

Table 2.—Minimum distance between components of grey water disposal system

Component of System	Horizontal Distance (feet)				
	Well or suction line	Water Supply line (pressure)	Stream	Dwelling	Property line
Building sewer	50	10 (a)	50	--	--
Treatment tank	50	10	50	5	10
Disposal trench and Disposal Bed	100	25	50	20	5
Disposal Pit	100	50	50	20	10

(a) Where the water supply line must cross a sewer line, the bottom of the water service within 10 feet of the point of crossing, shall be at least 12 inches above the top of the sewer line. The sewer line shall be of cast iron with leaded or mechanical joints at least 10 feet on either side of the crossing.

Disposal Trenches

A field of disposal trenches consists of a field of 12 inch lengths of 4 inch agricultural drain tile, 2 to 3 foot lengths of vitrified clay sewer pipe, or perforated, nonmetallic pipe. In areas having unusual soil or water characteristics, local experience should be reviewed before selecting piping materials. The individual laterals preferably should not be over 100 feet long, and the trench bottom and tile distribution lines should be about level. Use of more and shorter laterals is preferred because if something should happen to disturb one line, most of the field will still be serviceable. From a theoretical moisture flow viewpoint, a spacing of twice the depth of gravel would prevent taxing the percolative capacity of the adjacent soil.

Many different designs may be used in laying out subsurface disposal fields. The choice may depend on the size and shape of the available disposal area, the capacity required, and the topography of the disposal area.

Typical layouts of disposal trenches are shown in Figures 4, 6, pages 11, and 16.

To provide the minimum required gravel depth and earth cover, the depth of the disposal trenches should be at least 24 inches. Additional depth may be needed for contour adjustment, extra aggregate under the tile, or other design purposes. The maintenance of a 4 foot separation between the bottom of the trench and the water table is required to minimize ground water contamination. In considering the depth of the disposal trenches, the possibility of tile lines freezing during prolonged cold period is raised. Freezing rarely occurs in a carefully constructed system kept in continuous operation. It is important during construction to assure that the tile lines are surrounded by gravel. Pipes under driveways or other surfaces which are usually cleared of snow should be insulated.

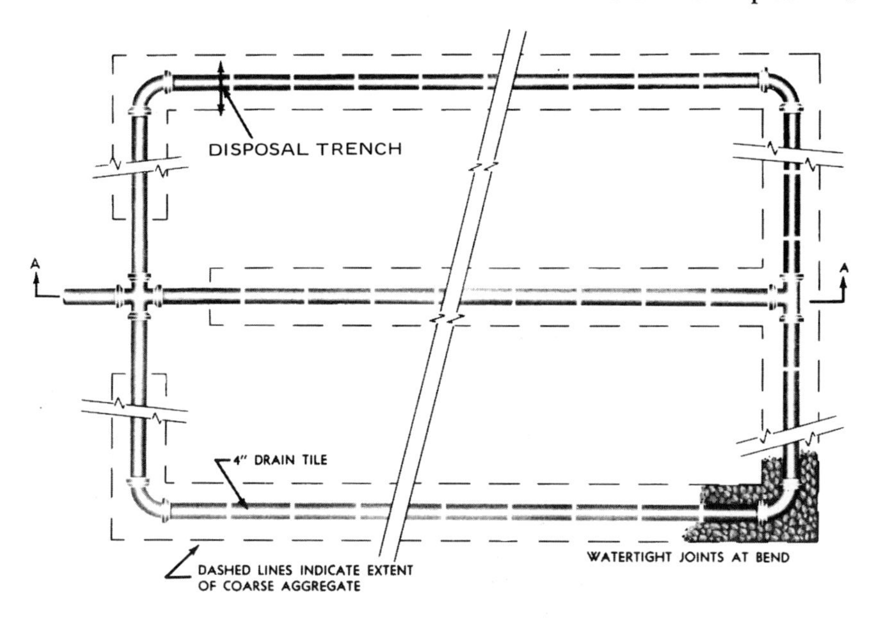

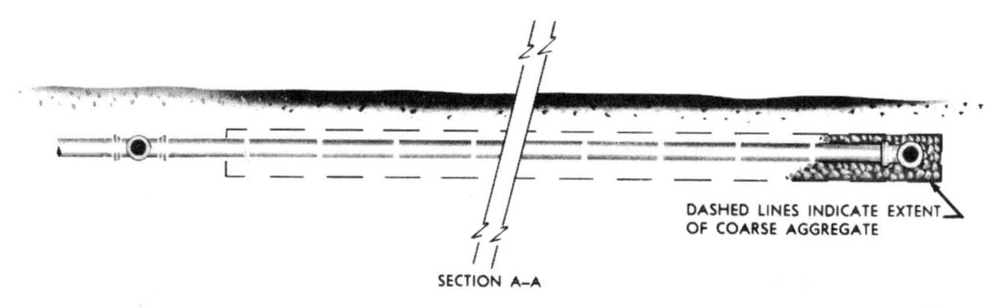

Figure 4.—Typical layout of disposal trench.

The required absorption area is predicated on the results of the soil precolation test, and may be obtained from column 2 or 4 of Table 1 (page 8), or Figure 3 (page 9). Note especially that the area requirements are per bedroom. The area of the lot on which the house is to be built should be large enough to allow room for an additional system if the first one fails. Thus for a 3 bedroom house on a lot where the minimum percolation rate was 1 inch in 15 minutes, the necessary absorption area will be 3 bedrooms x 115 sq. ft. per bedroom, or 345 sq. ft. For trenches 2 feet wide with 6 inches of gravel below the drain pipe, the required total length of the trench would be 345 ÷ 2, or 173 feet. If this were divided into 5 portions (i.e., 5 laterals), the length of each line would be 173 ÷ 5, or 35 feet. The spacing of trenches is generally governed by practical construction considerations dependent on the type of equipment, safety, etc. For serial distribution on sloping ground, trenches should be separated by 6 feet to prevent short circuiting. Table 2, page 10, gives the various distances the system has to be kept away from wells, dwellings, etc.

In the example cited, trenches are 2 feet wide x 5 trenches = 10 feet plus 6 feet between trenches x 4 spaces = 24 feet. The total width of 34 feet x 57 feet in length = 1,938 square feet, plus additional land required to keep the field away from wells, property lines, etc.

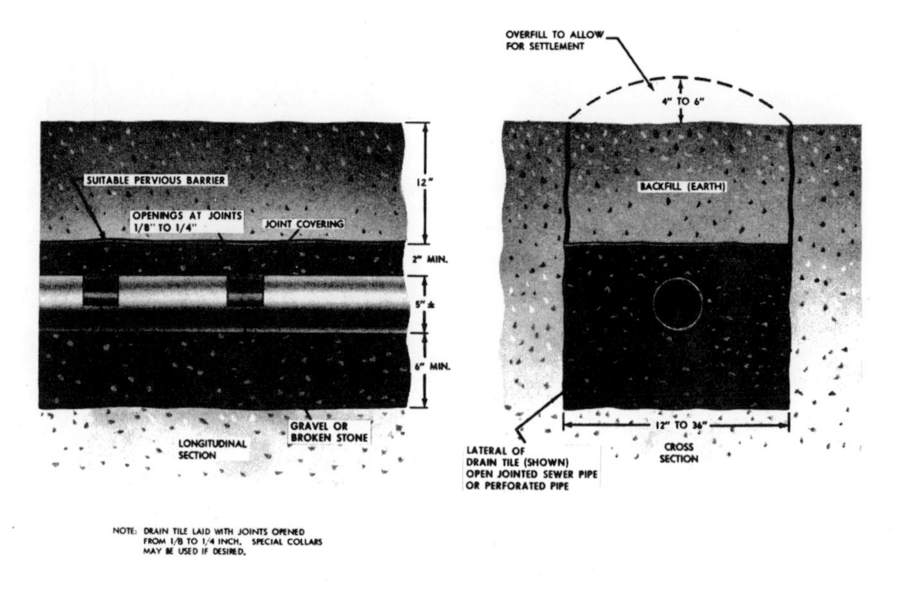

Figure 5.—Disposal trench and lateral.

Construction Considerations.—Careful construction is important in obtaining a satisfactory soil absorption system. Attention should be given to the protection of the natural absorption properties of the soil. Care must be taken to prevent sealing of the surface on the bottom and sides of the trench. Trenches should not be excavated when the soil is wet enough to smear or compact easily. Soil moisture is right for safe working only when a handful will mold with considerable pressure. Open trenches should be protected from surface runoff to prevent the entrance of silt and debris. If it is necessary to walk in the trench, a temporary board laid on the bottom wili reduce the damage. Some smearing and damage is bound to occur. All smeared or compacted surfaces should be raked to a depth of 1 inch, and loose material removed, before the gravel is placed in the trench.

The pipe, laid in a trench of sufficient width and depth, should be surrounded by clean, graded gravel or rock, broken hard burned clay brick, or similar aggregate. The material may range in size from ½ inch to 2½ inches. Cinders, broken shell, and similar material are not recommended, because they are usually too fine and may lead to premature clogging. The material should extend from at least 2 inches above the top of the pipe to at least 6 inches below the bottom of the pipe. If tile is used, the upper half of the joint openings should be covered, as shown in Figure 5. The top of the stone should be covered, with untreated building paper, a 2-inch layer of hay or straw, or similar pervious material to prevent the stone from becoming clogged by the earth backfill. An impervious covering should not be used, as this interferes with evaportranspiration at the surface (see Appendix A, page 37). Although generally not figured in the calculations, evapotranspiration is often an important factor in the operation of horizontal absorption systems.

Drain tile connectors, collars, clips, or other spacers with covers for the upper half of the joints are of value in obtaining uniform spacing, proper alignment, and protection of tile joints, but use of such aides is optional. They have been made of galvanized iron, copper, and plastic.

It has been found that root problems may be prevented best by using a liberal amount of gravel or stone around the distribution pipe. Clogging due to roots has occurred mostly in trenches with insufficient gravel around distribution pipes. Furthermore, roots seek the location where moisture conditions are most favorable for growth and, in the small percentage of cases where they become troublesome in well designed installations, there is usually some explanation involving the moisture conditions. At a residence which is used only during the summer, for example, roots are most likely to penetrate when the house is uninhabited, or when moisture immediately below or around the gravel becomes less plentiful than during the period when the system is in use. In general, trenches constructed within 10 feet of large trees or dense shrubbery should have at least 12 inches of gravel or crushed stone beneath the tile.

The top of a new disposal trench should be hand tamped and should be overfilled with about 4 to 6 inches of earth. Unless this is done, the top of the trench may settle to a point lower than the surface of the adjacent ground. This will cause the collection of storm waters in the trench, which can lead to premature saturation of the disposal field and possibly to complete washout of the trench. Machine tamping or hydraulic backfilling of the trench should be prohibited.

Where sloping ground is used for the disposal area, it is usually necessary to construct a small temporary dike or surface water diversion ditch above the field, to prevent the disposal area from being washed out by rain. The dike should be maintained or the ditch kept free of obstructions until the field becomes well covered with vegetation.

A heavy vehicle would readily crush the tile in a shallow disposal field. For this reason, heavy machinery should be excluded from the disposal area unless special provision is made to support the weight. All machine grading should be completed before the field is laid.

The use of the field area must be restricted to activities which will not contribute to the compaction of the soil with the consequent reduction in soil aeration.

Disposal Beds

Common design practice for disposal fields for private residences provides for trench widths up to 36 inches. Variations of design utilizing increased widths are being used in many areas. Disposal fields having trenches wider than 3 feet are referred to as disposal beds. The design of trenches is based on an empirical relationship between the percolation test and the bottom area of the trenches. The use of disposal beds has been limited by the lack of experience with their performance and the absence of design criteria comparable to those for trenches.

Studies sponsored by the Federal Housing Administration have demonstrated that the disposal bed is a satisfactory device for disposing of effluent in soils that are acceptable for soil absorption systems. The studies have further demonstrated that the empirical relationship between the percolation test and bottom area required for trenches is applicable to beds.

There are three main elements of a disposal bed: absorption surface, rockfill or packing material, and the distribution system. The design of the bed should be such that the total intended absorption area is preserved, sufficient packing material is provided in the proper place to allow for further treatment and storage of excess liquid, and a means for distributing the effluent is protected against siltation of earth backfill and mechanical damage. Construction details for a conventional disposal bed are outlined in the following material in such a way that these principal design elements are incorporated. Tabulation of construction details for the conventional disposal bed is not intended to preclude other designs which may provide the essential features in a more economical or otherwise desirable manner. Specifically, there may be equally acceptable or even superior methods developed for distributing the liquid than by tile or perforated pipe covered with gravel.

The use of disposal beds results in the following advantages:

1. A wide bed makes more efficient use of land available for disposal fields than a series of long narrow trenches with wasted land between the trenches.

2. Efficient use may be made of a variety of modern earth moving equipment employed at housing projects for other purposes such as basement excavation and landscaping, resulting in savings on the cost of the system.

Construction Considerations.—When disposal beds are used, the following design and construction procedures providing for rockfill or packing material, an adequate distribution system, and protection of the absorption area, should be observed:

1. The amount of bottom absorption area required shall be the same as shown in Table 1, page 8.

2. Percolation tests should be conducted in accordance with pages 4-6.

3. The bed should have a minimum depth of 24 inches below natural ground level to provide a minimum earth backfill cover of 12 inches.

4. The bed should have a minimum depth of 12 inches of rockfill or packing material extending at least 2 inches above and 6 inches below the distribution pipe.

5. The bottom of the bed and distribution tile or perforated pipe should be level.

6. Lines for distributing effluent shall be spaced not greater than 6 feet apart and not greater than 3 feet from the bed sidewall.

7. When more than one bed is used: (a) there should be a minimum of 6 feet of undisturbed earth between adjacent beds; and (b) the beds should be connected in series in accordance with the section concerning serial distribution, below.

8. Applicable construction considerations for standard trenches on pages 12-13 should also be followed.

Serial Distribution

Serial distribution is achieved by arranging individual components of the disposal field so that each component is forced to pond to the full depth of the gravel fill before liquid flows into the succeeding component.

Serial distribution has the following advantages:

1. Serial distribution minimizes the importance of variable absorption rates by forcing each component of a disposal field to absorb effluent until its ultimate capacity is utilized. The variability of soils even in the small area of an individual disposal field raises doubt of the desirability of uniform distribution. Any one or a combination of factors may lead to nonuniform absorptive capacity of the several components of a system. Varying physical and chemical characteristics of soil, construction damage such as soil interface smearing or excessive compaction, poor surface drainage, and variation in depth of trenches are some of the factors involved.

2. Serial distribution causes successive components in the disposal field to be used to full capacity. Serial distribution has a distinct advantage on sloping terrain. With imperfect division of flow in a parallel system, one trench could become overloaded, resulting in a surcharged condition. If the slope of the ground and elevation of the distribution box were such that a surcharged trench continued to re-receive more effluent than it could absorb, local failure would occur before the full capacity of the system was utilized.

3. The cost of the distribution box is eliminated in serial distribution. Also, long runs of closed pipe connecting the box to each trench are unnecessary.

Fields in Flat Areas.—Where the slope of the ground surface does not exceed six inches in any direction within the area utilized for the disposal field, the grey water treatment tank effluent may be applied to the disposal field through a system of interconnected distribution pipes in a continuous system of trenches. The following specific criteria should be followed:

1. A minimum of 12 inches of earth cover is provided over the gravel fill in all trenches of the system.

2. The bottom of the trenches and the distribution lines should be level.

3. One type of a satisfactory disposal field layout for "level" ground is shown in Figure 6, below.

4. Construction considerations for standard trenches, pages 12-13 should be followed.

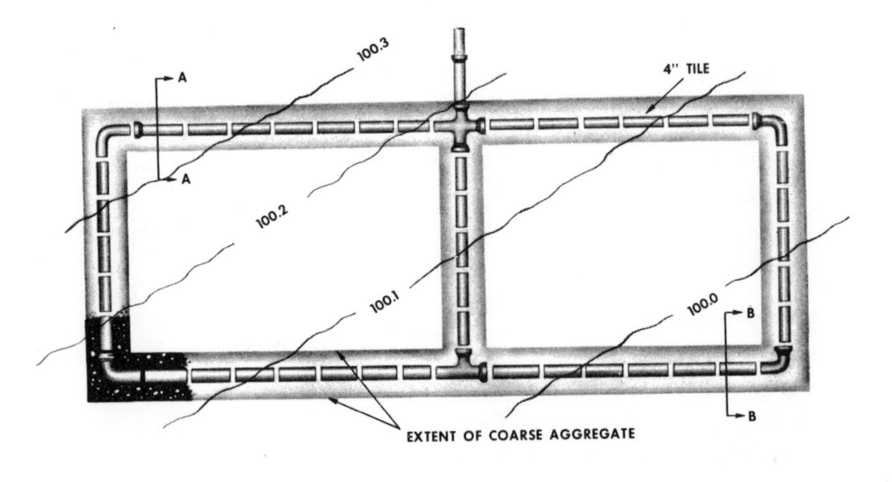

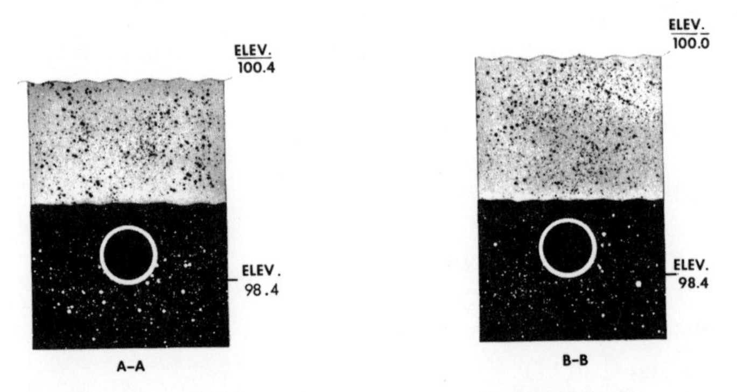

Figure 6.—Disposal-field system for level ground.

Fields in Sloping Ground.—Serial distribution may be used in all situations where a disposal field is permitted and should be used where the fall of the ground surface exceeds approximately 6 inches in any direction within the area utilized for the disposal field. The maximum ground slope suitable for serial distribution systems should be governed by local factors affecting the erosion of the ground used for the disposal field. Excessive slopes which are not protected from surface water runoff or do not have adequate vegetation cover to prevent erosion should be avoided. Generally, ground having a slope greater than one vertical to two horizontal should be investigated carefully to determine if satisfactory from the erosion standpoint. Also, the horizontal distance from side of the trench to the ground surface should be adequate to prevent lateral flow of effluent and breakout on surface and in no case be less than two feet.

In serial distribution, each adjacent trench (or pair of trenches) is connected to the next by a closed pipe line laid on an undisturbed section of ground, as shown in Figure 7, page 17. The arrangement is such that all effluent is

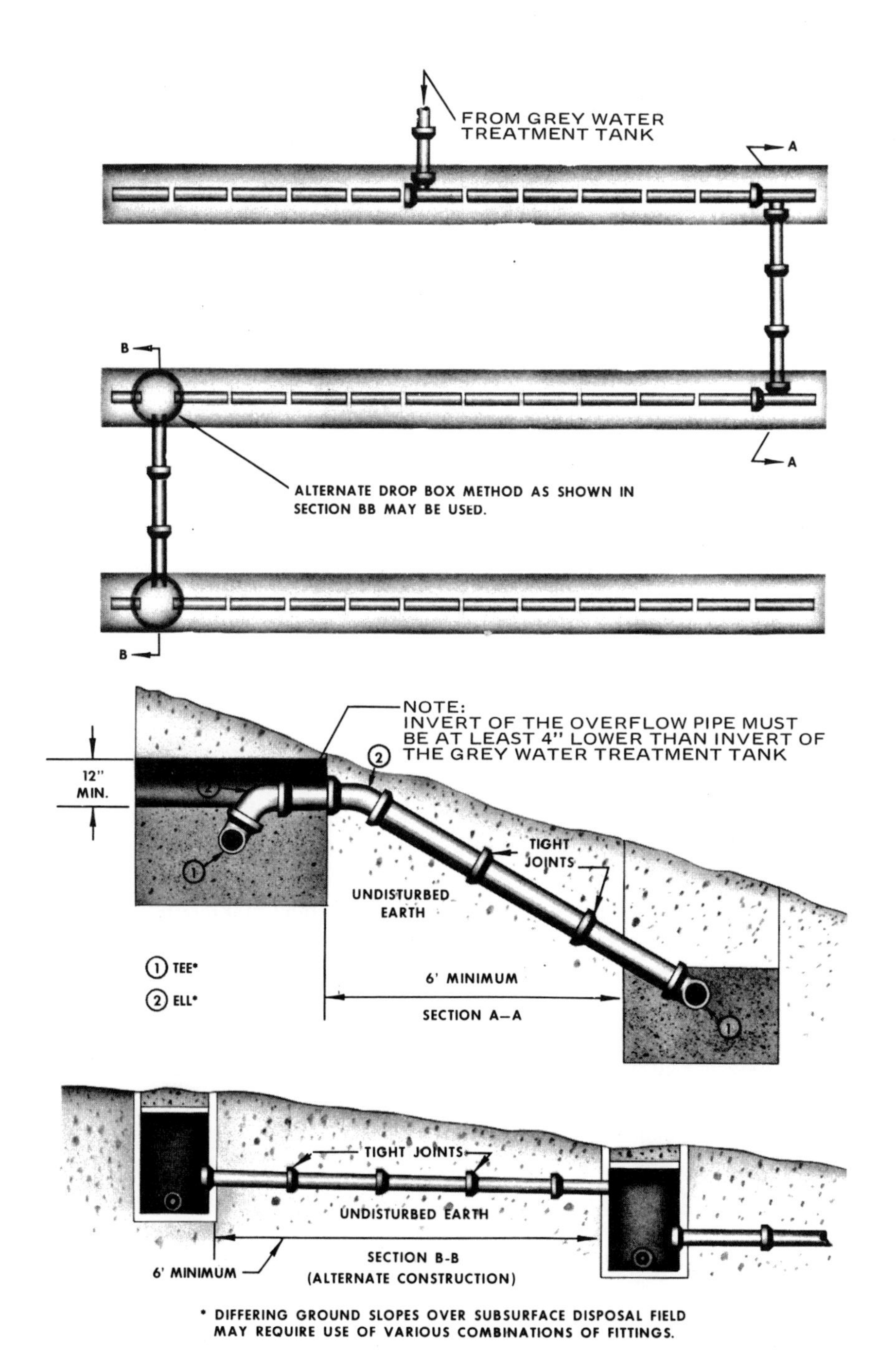

Figure 7.—A relief line arrangement for serial distribution.

discharged to the first trench until it is filled. Excess liquid is then carried by means of a closed line to the next succeeding or lower trench. In that manner, each portion of the subsurface system is used in succession. When serial distribution is used, the following design and construction procedures should be followed:

1. The bottom of each trench and its distribution line should be level.
2. There should be a minimum of 12 inches of ground cover over the gravel fill in the trenches.
3. The disposal trenches should follow approximately the ground surface contours so that variations in trench depth will be minimized.
4. There should be a minimum of 6 feet of undisturbed earth between adjacent trenches and between the grey water treatment tank and the nearest trench.
5. Adjacent trenches may be connected with the relief line or a drop box arrangement, Figure 7, page 17, in such a manner that each trench is completely filled with treated grey water to the full depth of the gravel before it flows to succeeding trenches. (The Figure shown does not preclude the use of other arrangements to provide serial distribution.)
    a. Trench connecting lines should be 4 inch, tight-joint sewers with direct connections to the distribution lines in adjacent trenches or to a drop box arrangement.
    b. Care must be exercised in constructing relief lines to insure an undisturbed block of earth between trenches. The trench for the relief pipe, where it connects with the preceding disposal trench, shall be dug no deeper than the top of the gravel. The relief line should rest on undisturbed earth and backfill should be carefully tamped.
    c. The relief lines connecting individual trenches should be as far from each other as practicable in order to prevent short circuiting.
6. Invert of the overflow pipe in the first relief line must be at least 4 inches lower than the invert of the grey water treatment tank outlet, Figure 7.
7. All other construction features of the disposal field are the same as recommended on pages 12-13.

Deep Disposal Trenches and Disposal Beds

In cases where the depth of filter material below the distribution pipe exceeds the standard six inch depth, credit may be given for the added absorption area provided in deeper trenches with a resultant decrease in length of trench. Such credit shall be given in accordance with Table 3 which gives the percentage of length of standard disposal trench (as computed from Table 1), based on six inch increments of increase in depth of filter material.

Table 3.—Percentage of length of standard trench[1]

Depth of Gravel Below Pipe in Inches[2]	Trench width 12"	Trench width 18"	Trench width 24"	Trench width 36"	Trench width 48"	Trench width 60"
12..................	75	78	80	83	86	87
18..................	60	64	66	71	75	78
24..................	50	54	57	62	66	70
30..................	43	47	50	55	60	64
36..................	37	41	44	50	54	58
42..................	33	37	40	45	50	54

[1]The standard disposal trench is one in which the filter material extends two inches above and six inches below the pipe.

[2]For trenches or beds having width not shown in Table 3, the percent of length of standard trench may be computed as follows:

$$\text{Percent of length standard trench} = \frac{w + 2}{w + 1 + 2d} \times 100$$

Where w = width of trench in feet
d = depth of gravel below pipe in feet

To use this table, consider the example on page 11. Using a trench 2 feet wide with 6" of gravel under tile, 173 feet are required. If the depth of gravel is increased to 18", keeping trench width at 2 feet, only 66% of 173 feet is required, or 114 feet. If 2 laterals are used, the length would be 114 divided by 2 = 57 feet.

The space between lines for serial distribution on sloping ground is 6 feet plus 2 lines x 2 feet = 4 feet. Total land required is 10 feet in width x 57 feet in length = 570 square feet, plus additional area required to keep the field away from wells, property lines, etc.

Disposal Pits

Disposal pits, as with all soil absorption systems, should never be used where there is a likelihood of contaminating underground waters, nor where adequate disposal beds or trenches can be provided. When disposal pits are to be used, the pit excavation should terminate 4 feet above the ground water table.

In some States, disposal pits are permitted as an alternative when shallower disposal fields are impracticable, and where the top 3 or 4 feet of soil is underlaid with porous sand or fine gravel and the subsurface conditions are otherwise suitable for pit installations. Where circumstances permit, pits may be either supplemental or alternative to the more shallow disposal fields. When pits are used in combination with shallower fields, the absorption areas in each system should be pro-rated, or based upon the weighted average of the results of the percolation tests.

It is important that the capacity of a disposal pit be computed on the basis of percolation tests made in each vertical stratum penetrated. The weighted average of the results should be computed to obtain a design figure. Soil strata in which the percolation rates are in excess of 30 minutes per inch should not

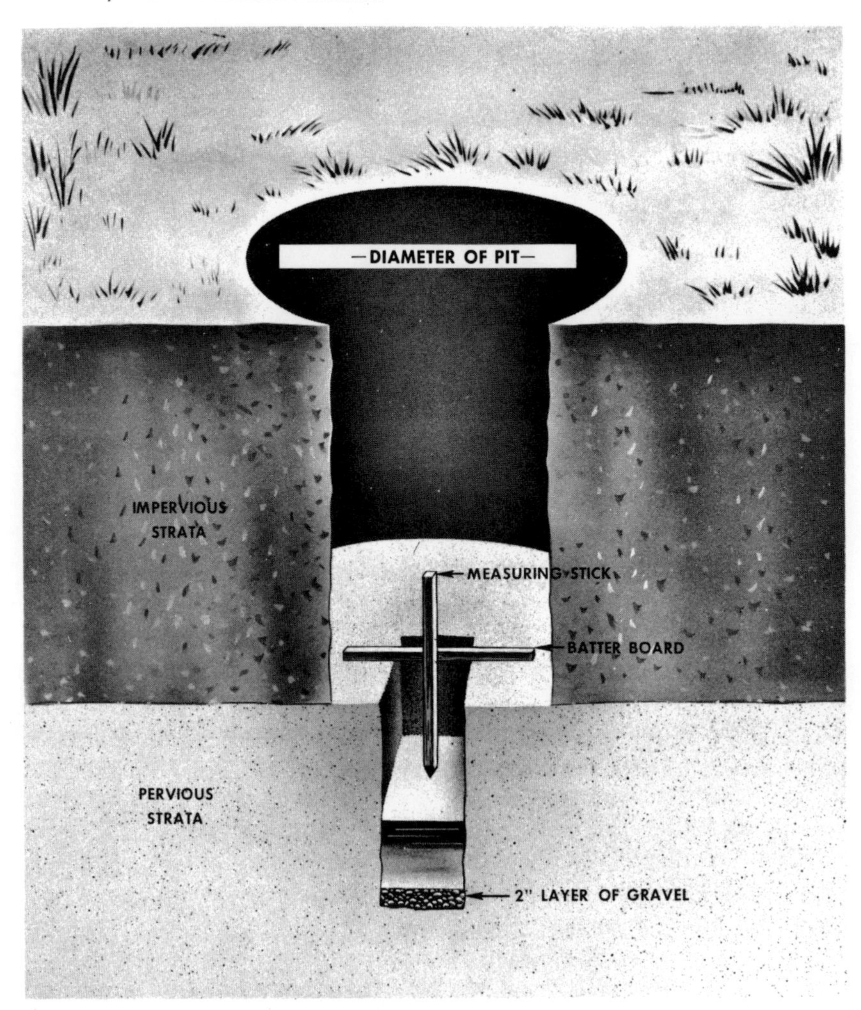

Figure 8.—Deep percolation test for disposal pit.

be included in computing the absorption area. As will be apparent from Figure 8 (above), adequate tests for deep pits are somewhat difficult to make, time-consuming, and expensive. Although few data have been collected comparing percolation test results with deep pit performance, nevertheless the results of such percolation tests, while of limited value, combined with competent engineering judgment based on experience, are the best means of arriving at design data for disposal pits.

Table 1 or Figure 3 (page 8) gives the absorption area requirements per gallon of grey water per day for the percolation rate obtained. The effective area of the disposal pit is the vertical wall area (based on dug diameter) of the pervious strata below the inlet. No allowance should be made for impervious strata or bottom area. With this in mind, Table 4 may be used for determining the effective side-wall area of circular or cylindrical disposal pits.

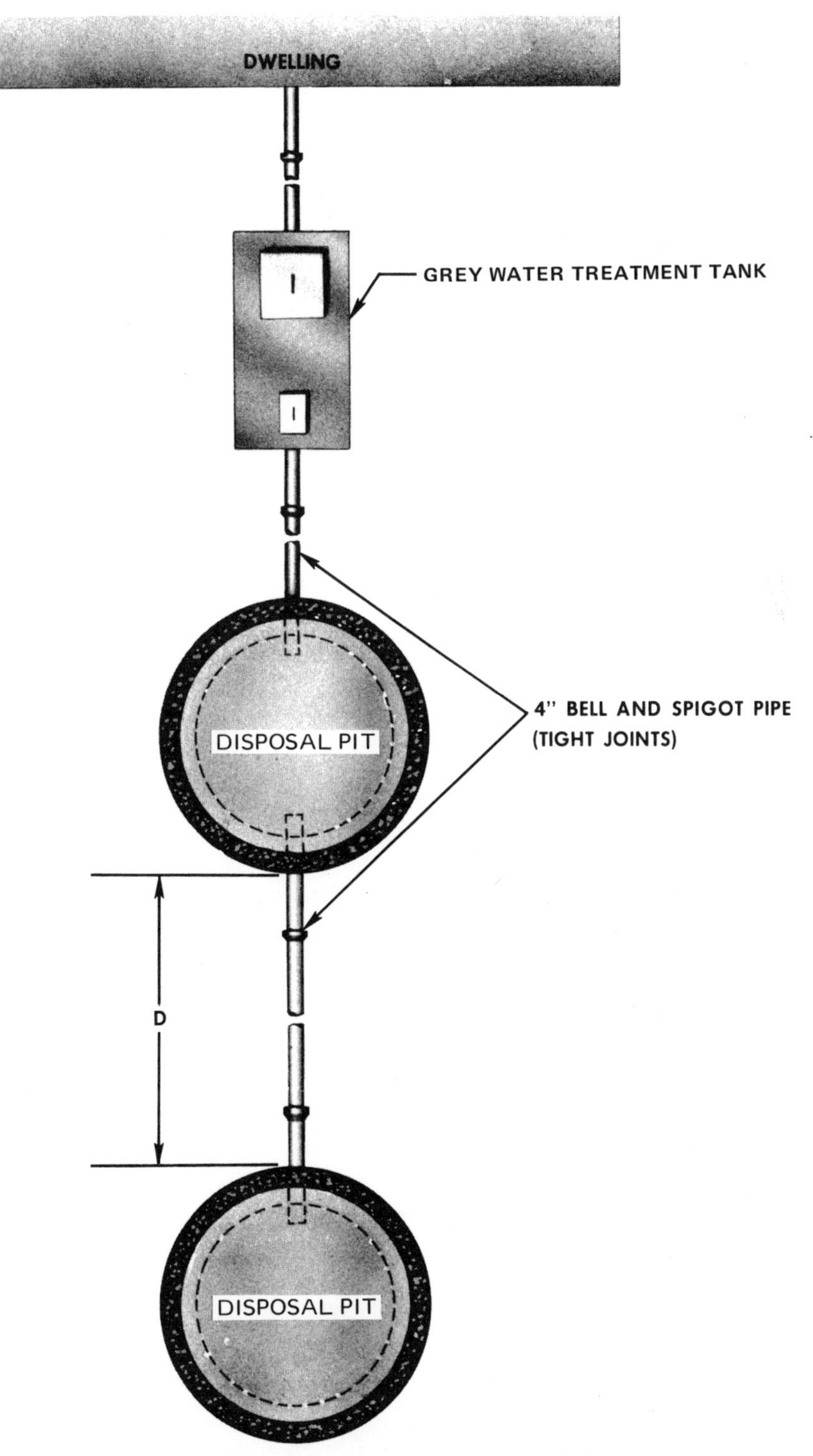

Figure 9.—Disposal system using two disposal pits.

Table 4.—Vertical wall areas of circular disposal pits

[In Square Feet]

Diameter of disposal pit (feet)	Effective strata depth below flow line (below inlet)									
	1 foot	2 feet	3 feet	4 feet	5 feet	6 feet	7 feet	8 feet	9 feet	10 feet
3.............	9.4	19	28	38	47	57	66	75	85	94
4.............	12.6	25	38	50	63	75	88	101	113	126
5.............	15.7	31	47	63	79	94	110	126	141	157
6.............	18.8	38	57	75	94	113	132	151	170	188
7.............	22.0	44	66	88	110	132	154	176	198	220
8.............	25.1	50	75	101	126	151	176	201	226	251
9.............	28.3	57	85	113	141	170	198	226	254	283
10.............	31.4	63	94	126	157	188	220	251	283	314
11.............	34.6	69	104	138	173	207	242	276	311	346
12.............	37.7	75	113	151	188	226	264	302	339	377

Example: A pit of 5 foot diameter and 6 foot depth below the inlet has an effective area of 94 square feet. A pit of 5 foot diameter and 16 foot depth has an area of 94 + 157, or 251 square feet.

Sample Calculations.—Assume that a disposal pit system is to be designed for a 3-bedroom home on a lot where the minimum percolation rate of 1 inch in 15 minutes prevails. According to Table 1, 3 x 115 (or 345) square feet of absorption area would be needed. Assume also that the water table does not rise above 27 feet below the ground surface, that disposal pits with effective depth of 20 feet can be provided, and that the house is in a locality where it is common practice to install disposal pits of 5-feet diameter (i.e., 4 feet to the outside walls, which are surrounded by about 6 inches of gravel). Design of the system is as follows:

Let d = depth of pit in feet; D = pit diameter in feet:

$$\pi Dd = 345 \text{ square feet}$$

$$3.14 \times 5 \times d = 345 \text{ square feet}$$

Solving for d depth of pit = 22 feet (approx.)

In other words, one 5 foot diameter pit 22 feet deep would be needed, but since the maximum effective depth is 20 feet in this particular location, it will be necessary to increase the diameter of the pit, or increase the number of pits, or increase both of these. This is illustrated in the example below:

(a) Design for 2 pits with a 10 foot diameter; d = depth of each pit.

$$2 \times 3.14 \times 10 \times d = 345 \text{ square feet}$$

$$d = 5.5 \text{ feet deep}$$

Use 2 pits 10 feet in diameter and 5.5 feet deep.

(b) Design for 2 pits with a 5 foot diameter; d = depth of each pit.

$$2 \times 3.14 \times 5 \times d = 345 \text{ square feet}$$

$$d = 11 \text{ feet (approximately)}$$

Use 2 pits 5 feet in diameter and 11 feet deep.

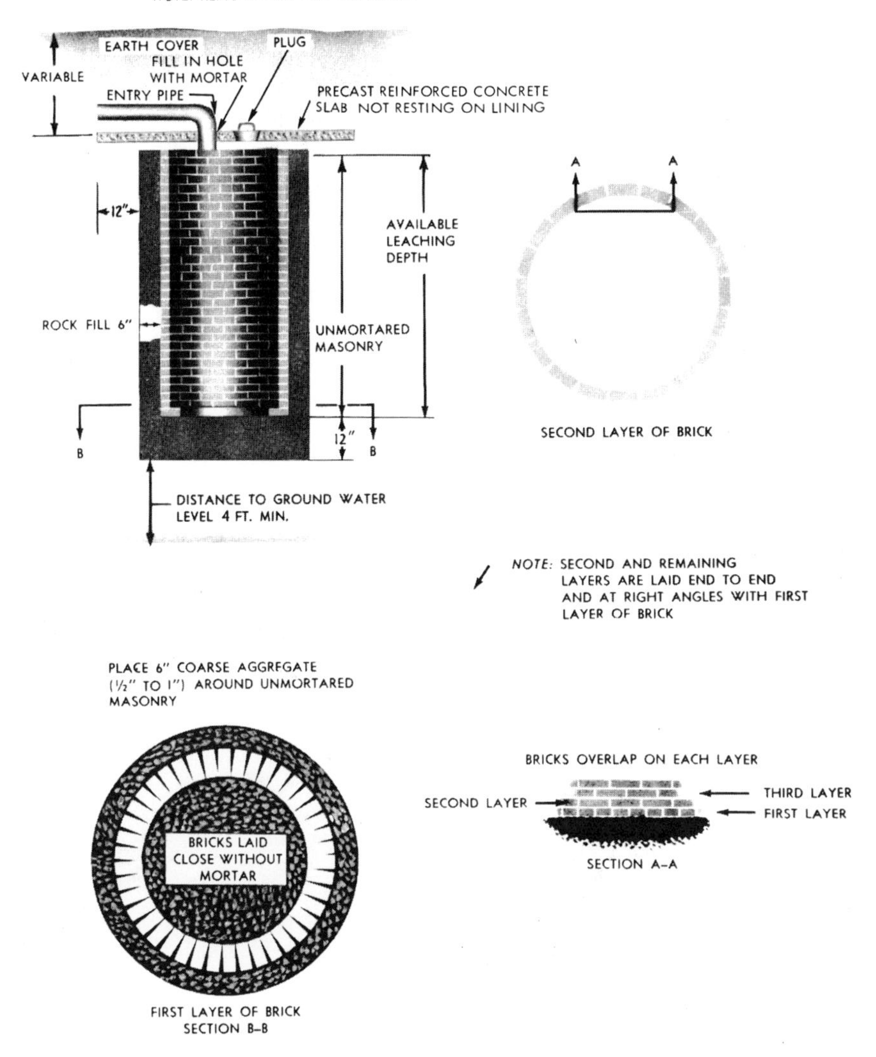

Figure 10.—Disposal pit.

Experience has shown that disposal pits should be separated by a distance equal to 3 times the diameter of the largest pit. For pits over 20 feet in depth, the minimum space between pits should be 20 feet (See Fig. 9, page 21). The area of the lot on which the house is to be built should be large enough to maintain this distance between the pits while still allowing room for additional pits if the first ones should fail. If this can be done, such a disposal field may be approved; if not, other suitable treatment and disposal facilities should be required.

Construction Considerations.—Soil is susceptible to damage during excavation. Digging wet soils should be avoided as much as possible. Cutting teeth on

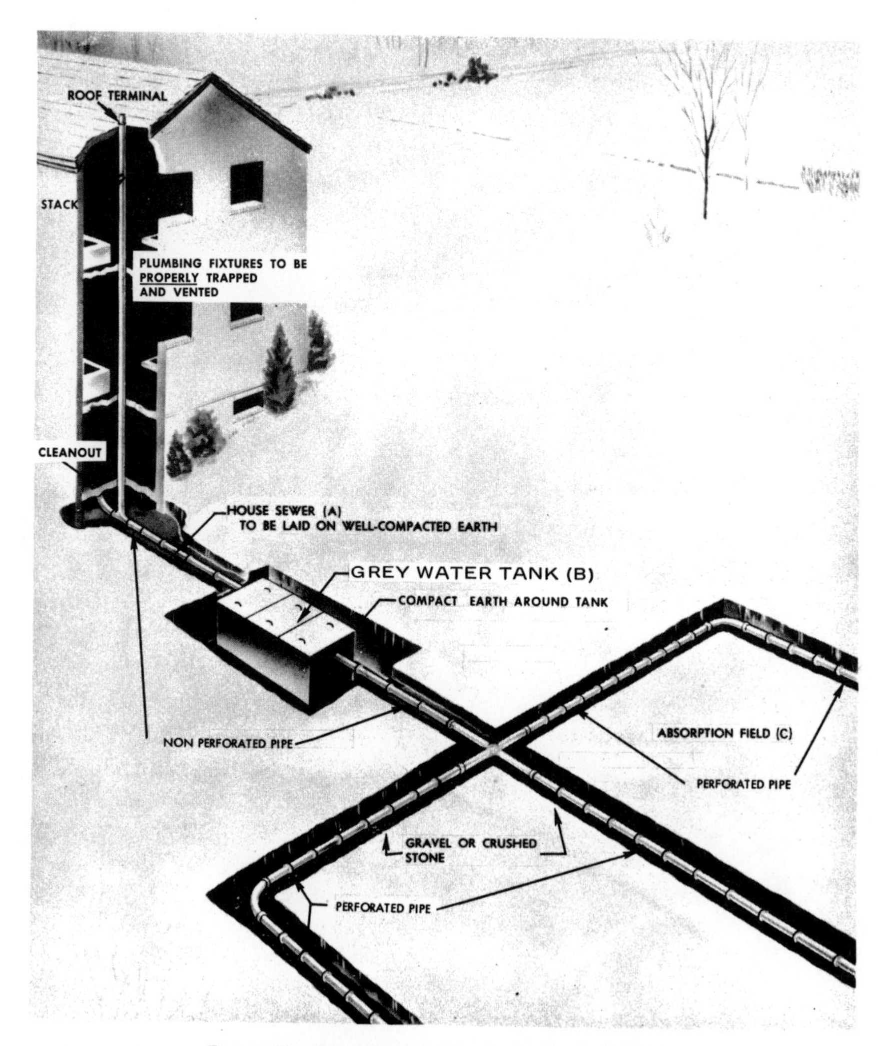

Figure 11.–Grey water treatment/disposal system.

mechanical equipment should be kept sharp. Bucket augered pits should be reamed to a larger diameter than the bucket. All loose material should be removed from the excavation.

Pits should be backfilled with clean gravel to a depth of one foot above the pit bottom or one foot above the reamed ledge to provide a sound foundation for the lining. Preferred lining materials are clay or concrete brick, block, or rings. Rings should have weep holes or notches to provide for seepage. Brick and block should be laid dry with staggered joints. Standard brick should be laid flat to form a four-inch wall. The outside diameter of the lining should be at least six inches less than the least excavation diameter. The annular space formed should be filled with clean, coarse gravel to the top of the lining as shown in Figure 10.

Either brick dome or flat concrete covers are satisfactory. They should be based on undisturbed earth and extend at least 12 inches beyond the excavation and should not bear on the lining for structural support. Bricks should be either laid in cement mortar or have a two-inch covering of concrete. If flat covers are used, a prefabricated type is preferred, and they should be reinforced to adequate strength. A nine inch capped opening in the pit cover is convenient for pit inspection. All concrete surfaces should be coated with a protective bitumastic or similar compound to minimize corrosion.

Connecting lines should be of a sound, durable material the same as used for the house to a grey water treatment tank connection. All connecting lines should be laid on a firm bed of undisturbed soil throughout their length. The grade of a connecting line should be at least two percent. The pit inlet pipe should extend horizontally at least one foot into the pit with a tee or ell to divert flow downward to prevent washing and eroding of the sidewalls. If multiple pits are used, or in the event repair pits are added to an existing system, they should be connected in series.

Abandoned disposal pits should be filled with earth or rock.

SELECTION OF A GREY WATER TREATMENT TANK

Assuming that the lot will be large enough to accommodate one of the types of disposal systems, and that construction of the system is permitted by local authority, the next step will be selection of a suitable grey water treatment tank.

Functions of Grey Water Treatment Tanks

Untreated liquid household grey waters can quickly clog soils. The tank conditions grey waters so that they may be more readily infiltrated into the subsoil of the ground. Thus, the most important function of a grey water treatment tank is to provide protection of the infiltrative capacity of the subsoil. Three functions take place within the tank to provide this protection.

1. *Removal of Solids.*–Clogging of soil with tank effluent varies directly with the amount of suspended solids in the liquid. As grey water from a building sewer enters a treatment tank, its rate of flow is reduced so that larger solids sink to the bottom or rise to the surface. These solids are retained in the tank, and the clarified effluent is discharged.

2. *Biological Treatment.*–Solids and liquid in the tank are subjected to decomposition by bacterial and natural processes. Bacteria present are of a variety called anaerobic which thrive in the absence of free oxygen. This decomposition or treatment of sewage under anaerobic conditions is termed "septic." Sewage which has been subjected to such treatment causes less clogging than untreated sewage containing the same amount of suspended solids.

3. *Sludge and Scum Storage.*–Sludge is an accumulation of solids at the bottom of the tank, while scum is a partially submerged mat of floating solids

that may form at the surface of the fluid in the tank. Sludge, and scum to a lesser degree, will be digested and compacted into a smaller volume. However, no matter how efficient the process is, a residual of inert solid material will remain. Space must be provided in the tank to store this residue during the interval between cleanings; otherwise, sludge and scum will eventually be scoured from the tank and may clog the disposal field.

If adequately designed, constructed, maintained, and operated, grey water treatment tanks are effective in accomplishing their purpose.

The relative position of a grey water treatment tank in a typical subsurface disposal system is illustrated in Figure 11 (page 24). The liquid contents of the house sewer (A) are discharged first into the grey water treatment tank (B), and finally into the subsurface disposal field (C).

The heavier sewage solids settle to the bottom of the tank, forming a blanket of sludge. The lighter solids, including fats and greases, rise to the surface and form a layer of scum. A considerable portion of the sludge and scum is liquefied through decomposition or digestion. During this process, gas is liberated from the sludge, carrying a portion of the solids to the surface, where they accumulate with the scum. Ordinarily, they undergo further digestion in the scum layer, and a portion settles again to the sludge blanket on the bottom. This action is retarded if there is much grease in the scum layer. The settling is also retarded because of gasification in the sludge blanket. Furthermore, there are relatively wider fluctuations of flow in small tanks than in the larger units. This effect has been recognized in Table 5 (page 27), which shows the recommended minimum liquid capacities of household grey water treatment tanks.

Location.—Grey water treatment tanks should be located where they cannot cause contamination of any well, spring, or other source of water supply. Underground contamination may travel in any direction and for considerable distances, unless filtered effectively. Underground pollution usually moves in the same general direction as the normal movement of the ground water in the locality. Ground water moves in the direction of the slope or gradient of the water table, i.e., from the area of higher water table to areas of lower water table. In general, the water table follows the general contour of the ground surface. For this reason, grey water treatment tanks should be located downhill from wells or springs. Sewage from disposal systems occasionally contaminates wells having higher surface elevations. Obviously, the elevations of disposal systems are almost always higher than the level of water in such wells as may be located nearby; hence, pollution from a disposal system on a lower surface elevation may still travel downward to the water bearing stratum as shown in Figure 12, below. It is necessary, therefore, to rely upon horizontal as well as vertical distances for protection. Tanks should never be closer than 50 feet from any source of water supply; and greater distances are preferred where possible.

The grey water treatment tank should not be located within 5 feet of any building, as structural damage may result during construction or seepage may enter the basement. The tank should not be located in swampy areas, or in

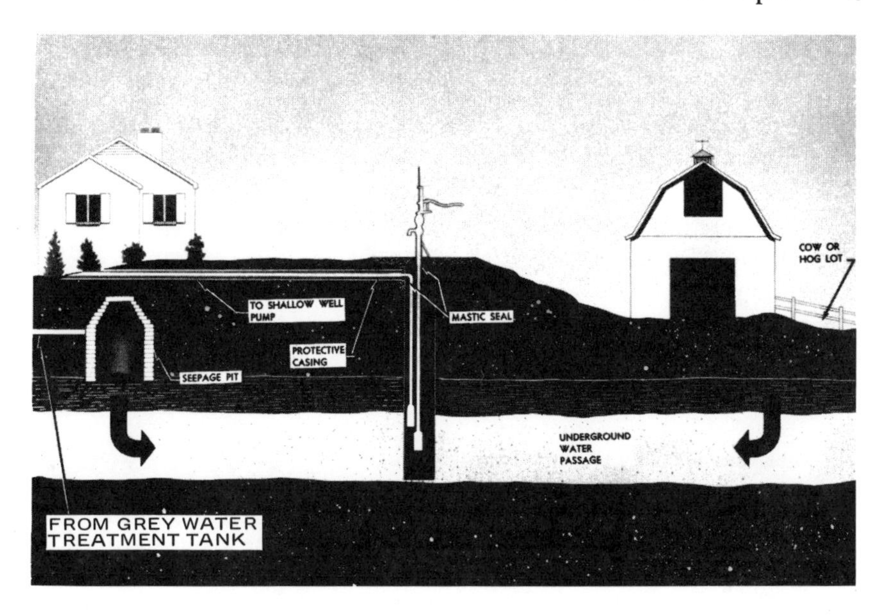

Figure 12.—Pollution of well from sources with lower surface elevations.

areas subject to flooding. In general, the tank should be located where the largest possible area will be available for the disposal field. Consideration should also be given to the location from the standpoint of cleaning and maintenance. Where public sewers may be installed at a future date, provision should be made in the household plumbing system for connection to such sewer.

Effluent.—Contrary to popular belief, grey water treatment tanks do not accomplish a high degree of bacteria removal. Although the sewage undergoes treatment in passing through the tank, this does not mean that infectious agents will be removed; hence, grey water treatment tank effluents cannot be considered safe. As previously explained, the primary purpose of the grey

Table 5.—Liquid capacity of grey water treatment tank (gallons)

[Provides for use of garbage grinders, automatic clothes washers, and other household appliances]

Number of bedrooms	Recommended minimum tank capacity	Equivalent capacity per bedroom
2 or less	450	225
3	540	180
4[1]	600	150

[1]For each additional bedroom, add 150 gallons.

water treatment tank is to condition the sewage so that it will cause less clogging of the disposal field.

Further treatment of the effluent, including the removal of pathogens, is effected by percolation through the soil. Disease producing bacteria will, in time, die out in the unfavorable environment afforded by soil. In addition, bacteria are also removed by certain physical forces during filtration. This combination of factors results in the eventual purification of the effluent.

Capacity.—Capacity is one of the most important considerations in grey water treatment tank design. Studies have proved that liberal tank capacity is not only important from a functional standpoint, but is also good economy. The liquid capacities recommended in Table 5 allow for the use of all household appliances, including garbage grinders.

Specifications for Grey Water Treatment Tanks

Materials.—Grey water treatment tanks should be watertight and constructed of materials not subject to excessive corrosion or decay, such as concrete, coated metal, vitrified clay, heavyweight concrete blocks, or hard burned bricks. Properly cured precast and cast-in-place reinforced concrete tanks are believed to be acceptable everywhere. Steel tanks meeting Commercial Standard 177-62 of the U.S. Department of Commerce are generally acceptable. Special attention should be given to job built tanks to insure water tightness. Heavyweight concrete block should be laid on a solid foundation and mortar joints should be well filled. The interior of the tank should be surfaced with two ¼ inch thick coats of portland cement-sand plaster. Some typical grey water treatment tanks are illustrated in Figure 13 (page 29). Suggested specifications for watertight concrete are given in Appendix C.

Precast tanks should have a minimum wall thickness of 3 inches, and should be adequately reinforced to facilitate handling. When precast slabs are used as covers, they should be watertight, have a thickness of at least 3 inches, adequately reinforced. All concrete surfaces should be coated with a bitumastic or similar compound to minimize corrosion.

General.—Backfill around grey water treatment tanks should be made in thin layers thoroughly tamped in a manner that will not produce undue strain on the tank. Settlement of backfill may be done with the use of water, provided the material is thoroughly wetted from the bottom upwards and the tank is first filled with water to prevent floating.

Adequate access must be provided to each compartment of the tank for inspection and cleaning. Both the inlet and outlet devices should be accessible. Access should be provided to each compartment by means of either a removable cover or a 20 inch manhole in least dimension. Where the top of the tank is located more than 18 inches below the finished grade, manholes and inspection holes should extend to approximately 8 inches below the finished grade (see Figure 14, page 30), or can be extended to finished grade if a seal is provided to keep odors from escaping. In most instances, the extension can be made using clay or concrete pipe, but proper attention must be given to the

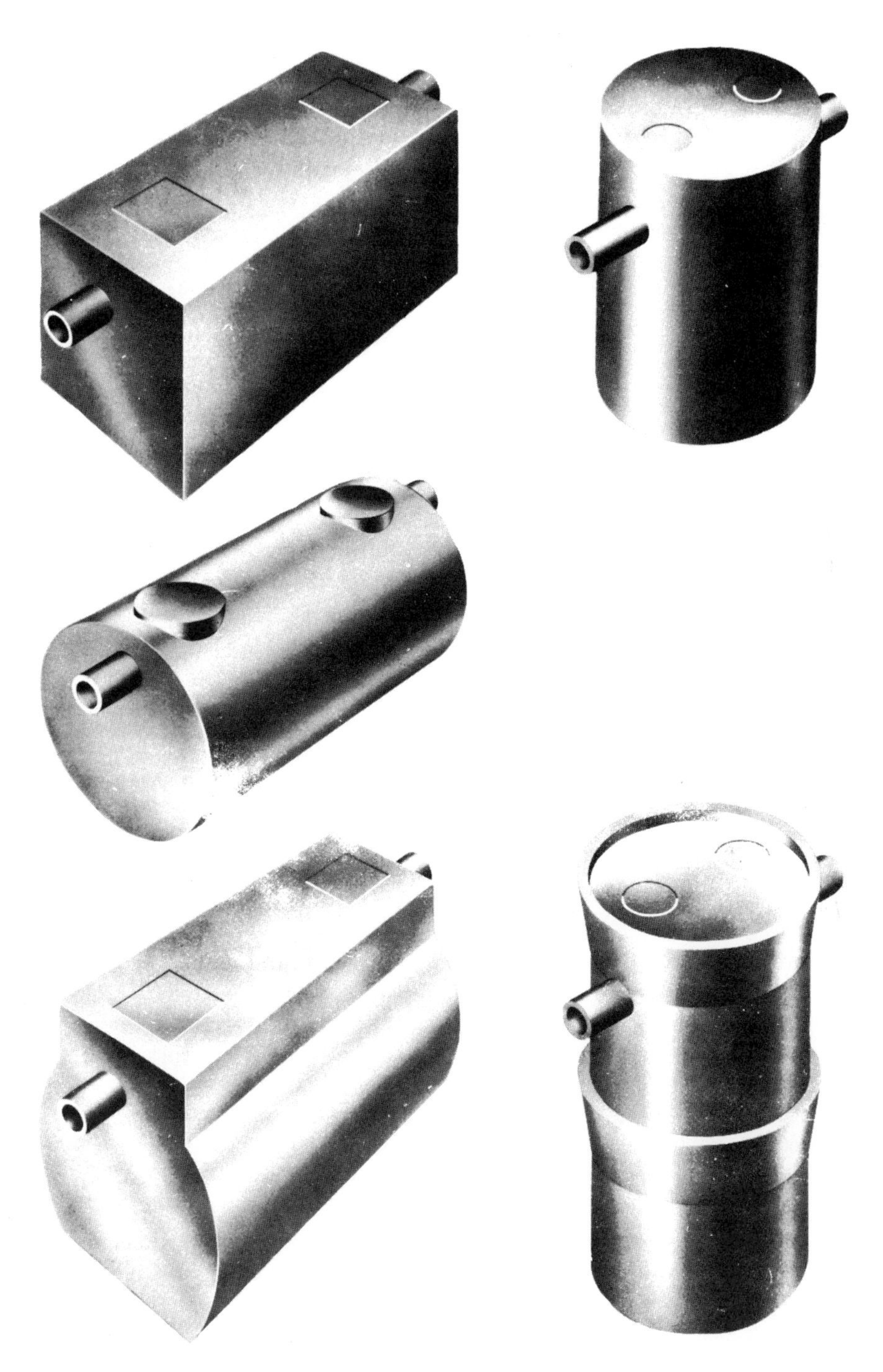

Figure 13.—Typical grey water treatment tank shapes

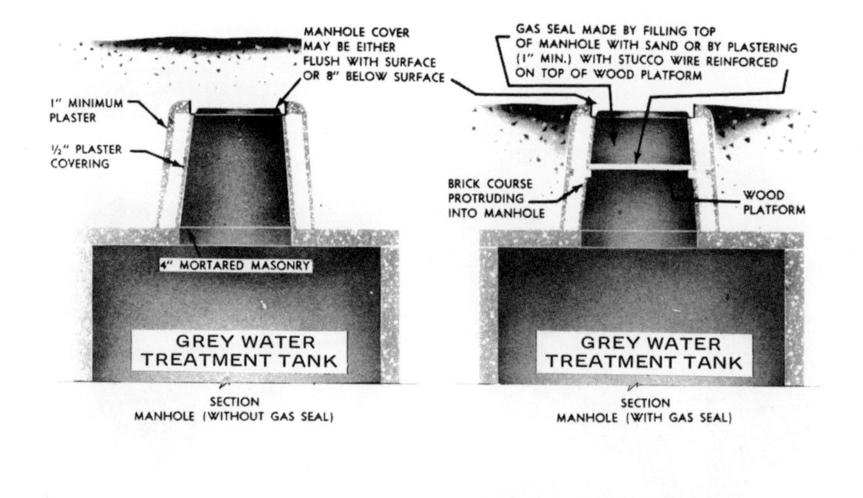

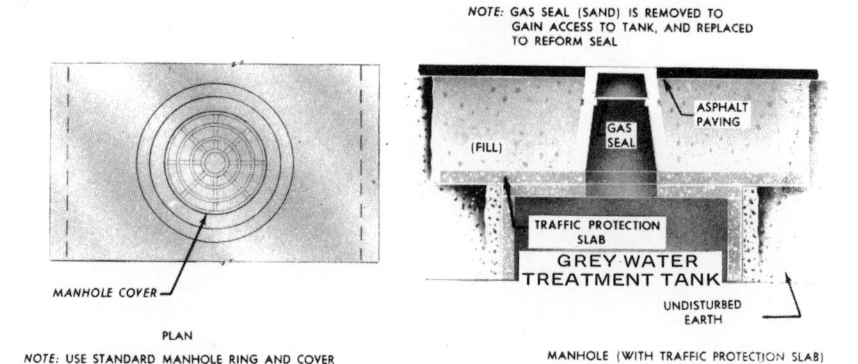

Figure 14.–Design of manholes.

accident hazard involved when manholes are extended close to the ground surface. Typical single and double compartment tanks are illustrated in Figures 15 and 17, pages 31 and 33.

Inlet.–The inlet invert should enter the tank at least 3 inches above the liquid level in the tank, to allow for momentary rise in liquid level during discharges to the tank. This free drop prevents backwater and stranding of solid material in the building sewer leading to the tank.

A vented inlet tee or baffle should be provided to divert the incoming grey waters downward. It should penetrate at least 6 inches below the liquid level, but in no case should the penetration be greater than that allowed for the outlet device. A number of arrangements commonly used for inlet and outlet devices are shown in Figure 16 (page 32).

Outlet.–It is important that the outlet device penetrate just far enough below the liquid level of the grey water treatment tank to provide a balance between sludge and scum storage volume; otherwise, part of the advantage of

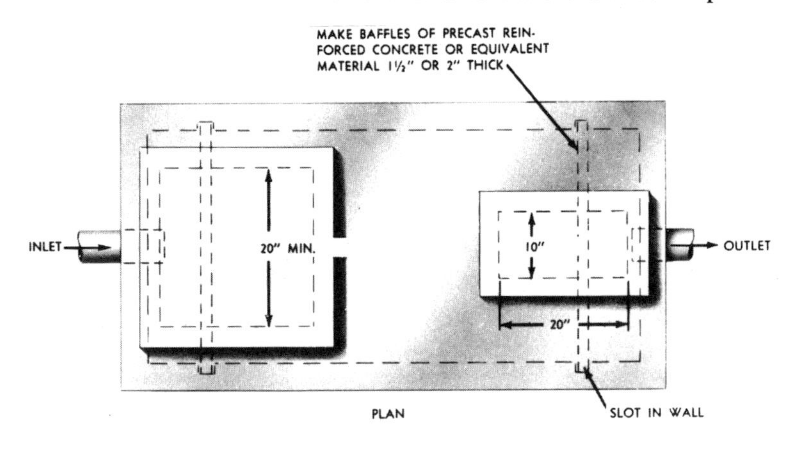

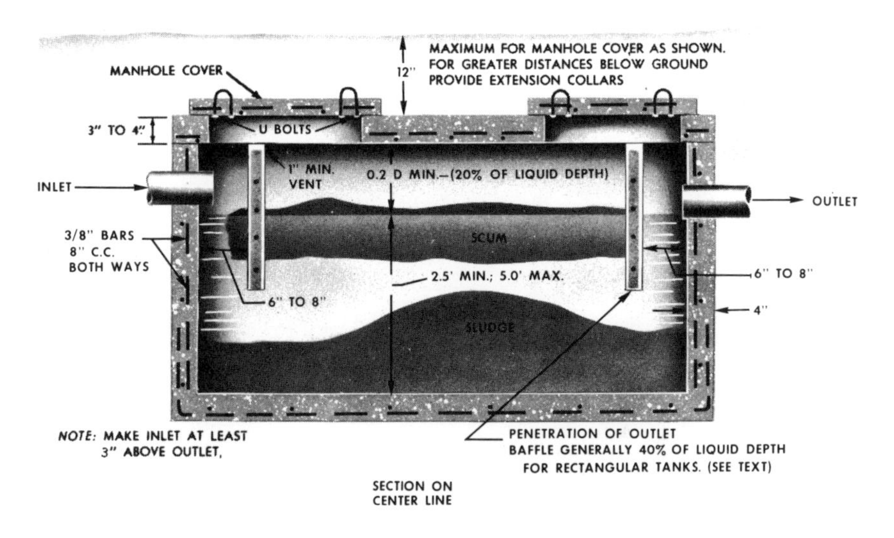

Figure 15.—Household grey water treatment tank.

capacity is lost. A vertical section of a properly operating tank would show it divided into three distinct layers; scum at the top, a middle zone free of solids (called "clear space"), and a bottom layer of sludge. The outlet device retains scum in the tank, but at the same time, it limits the amount of sludge that can be accommodated without scouring, which results in sludge discharging in the effluent from the tank. Observations of sludge accumulations in the field, as reported in Section B, Part III, of "studies on Household Sewage Disposal Systems" (bibliography reference, page 44), indicate that the outlet device should generally extend to a distance below the surface equal to 40 percent of the liquid depth. For horizontal, cylindrical tanks, this should be reduced to 35 percent. For example, in a horizontal cylindrical tank having a liquid depth of 42 inches, the outlet device should penetrate 42 X .35 = 14.7 inches below the liquid level.

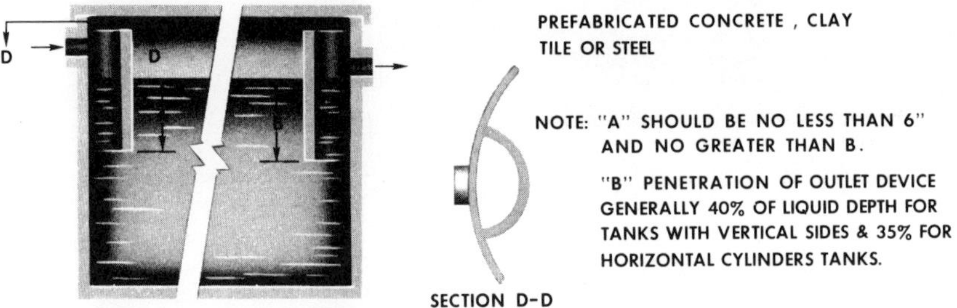

Figure 16.—Types of inlet and outlet devices.

The outlet device should extend above the liquid line to approximately one inch from the top of the tank. The space between the top of the tank and the baffle allows gas to pass off through the tank into the house vent.

Tank Proportions.—The available data indicate that, for tanks of a given capacity, shallow tanks function as well as deep ones. Also, for tanks of a given capacity and depth, the shape of a grey water treatment tank is unimportant. However, it is recommended that the smallest plan dimension be at least 2 feet. Liquid depth may range between 30 and 60 inches.

Storage Above Liquid Level.—Capacity is required above the liquid line to provide for that portion of the scum which floats above the liquid. Although some variation is to be expected, on the average about 30 percent of the total scum will accumulate above the liquid line. In addition to the provision for scum storage, one inch is usually provided at the top of the tank to permit free passage of gas back to the inlet and house vent pipe.

For tanks having straight, vertical sides, the distances between the top of the tank and the liquid line should be equal to approximately 20 percent of the liquid depth. In horizontal, cylindrical tanks, area equal to approximately 15 percent of the total circle should be provided above the liquid level. This condition is met if the liquid depth (distance from outlet invert to bottom of tank) is equal to 79 percent of the diameter of the tank.

Use of Compartments.—Although a number of arrangements are possible, compartments, as used here, refer to a number of units in series. These can be either separate units linked together, or sections enclosed in one continuous shell (as in Figure 17), with water-tight partitions separating the individual compartments.

A single compartment tank will give acceptable performance. The available research data indicate, however, that a two compartment tank, with the first compartment equal to one-half to two-thirds of the total volume, provides better suspended solids removal which may be especially valuable for

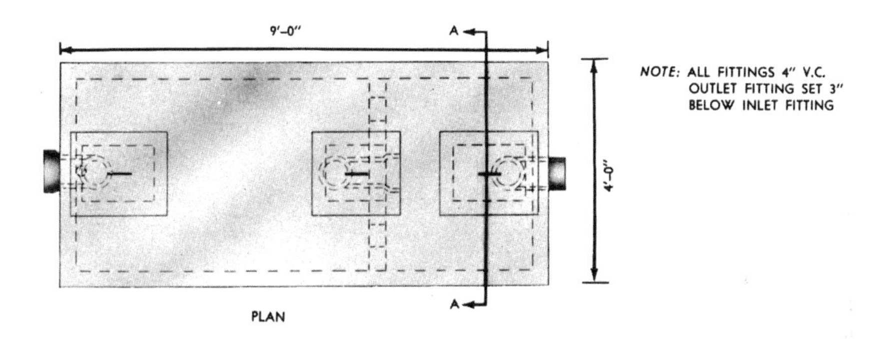

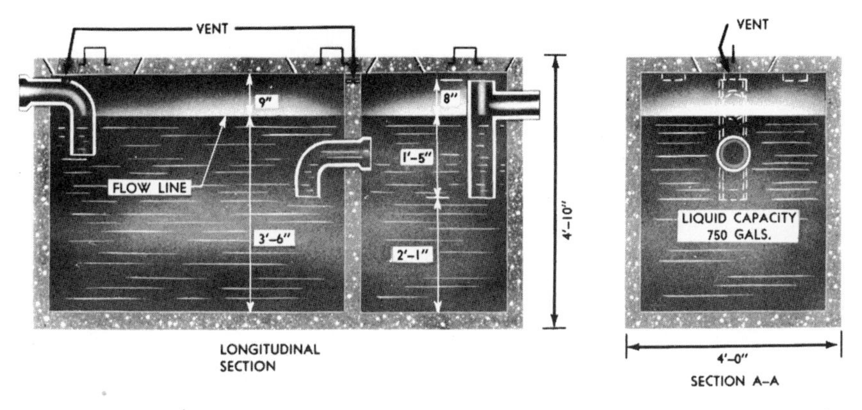

Figure 17.—Precast septic tank which could be used for a large grey water treatment tank.

protection of the soil absorption system. Tanks with three or more equal compartments give at least as good performance as single compartment tanks of the same total capacity. Each compartment should have a minimum plan dimension of 2 feet with a liquid depth ranging from 30 to 60 inches.

An access manhole should be provided to each compartment. Venting between compartments should be provided to allow free passage of gas. Inlet and outlet fittings in the compartmented tank should be proportioned as for a single tank. (See Figure 16, page 32.) The same allowance should be made for storage above the liquid line as in a single tank.

General Information on Grey Water Treatment Tanks

Cleaning.–Grey water treatment tanks should be cleaned before too much sludge or scum is allowed to accumulate. If either the sludge or scum approaches too closely to the bottom of the outlet device, particles will be scoured into the disposal field and will clog the system. Eventually, when this happens, liquid may break through to the ground surface, and/or the sewage may back up in the plumbing fixtures. When a disposal field is clogged in this manner, it is not only necessary to clean the tank, but it also may be necessary to construct a new disposal field.

The tank capacities given in Table 5 on page 27 will give a reasonable period of good operation before cleaning becomes necessary. There are wide differences in the rate that sludge and scum will accumulate from one tank to the next.

In most communities where septic tanks are used, there are firms which conduct a business of cleaning septic tanks. The local health department can make suggestions on how to obtain this service and frequency of cleaning needed for tanks typically installed in the locale. Firms servicing black water systems might also best service grey water treatment tanks. Cleaning is usually accomplished by pumping the contents of the tank into a tank truck. *Tanks should not be washed or disinfected after pumping.* A small residual of sludge should be left in the tank for seeding purposes. The material removed may be buried in uninhabited places or, with permission of the proper authority, emptied into a sanitary sewer system. It should never be emptied into storm drains or discharged directly into any stream or watercourse. Methods of disposal should be approved by the health authorities.

When a large grey water treatment tank is being cleaned, care should be taken not to enter the tank until it has been thoroughly ventilated and gases have been removed to prevent explosion hazards or asphyxiation of the workers. Anyone entering the tank should have one end of a stout rope tied around his waist, with the other end held above ground by another person strong enough to pull him out if he should be overcome by any gas remaining in the tank.

Grease Interceptors.–Grease interceptors (grease traps) are not ordinarily considered necessary on household sewage disposal systems. The discharge from a garbage grinder should never be passed through them. The grey water treatment tank capacities recommended in this manual are sufficient to receive the grease normally discharged from a home.

Chemicals.—The functional operation of grey water treatment tanks is not improved by the addition of disinfectants or other chemicals. In general, the addition of chemicals to a grey water treatment tank is not recommended. Some proprietary products which are claimed to "clean" grey water treatment tanks contain sodium hydroxide or potassium hydroxide as the active agent. Such compounds may result in sludge bulking and a large increase in alkalinity, and may interfere with digestion. The resulting effluent may severely damage soil structure and cause accelerated clogging, even though some temporary relief may be experienced immediately after application of the product.

Frequently, however, the harmful effects of ordinary household chemicals are overemphasized. Small amounts of chlorine bleaches, added ahead to the tank, may be used for odor control and will have no adverse effects. Small quantities of lye or caustics normally used in the home, added to plumbing fixtures is not objectionable as far as operation of the tank is concerned. If the grey water treatment tanks are as large as herein recommended, dilution of the lye or caustics in the tank will be enough to overcome any harmful effects that might otherwise occur.

Some 1,200 products, many containing enzymes, have been placed on the market for use in septic tanks, and extravagant claims have been made for some of them. As far as is known, however, none has been proved of advantage in properly controlled tests. It would not be expected that such additives would be beneficial to the grey water treatment tank.

Soaps, detergents, bleaches, drain cleaners, or other material, as normally used in the household, will have no appreciable adverse effect on the system. However, as both the soil and essential organisms might be susceptible to large doses of chemicals and disinfectants, moderation should be the rule. Advice of responsible officials should be sought before chemicals arising from a hobby or home industry are discharged into a grey water treatment/disposal system.

Miscellaneous.—Roof drains, foundation drains and drainage from other sources producing large intermittent or constant volumes of clear water should not be piped into the grey water treatment tank or disposal field area. Such large volumes of water will stir up the contents of the tank and carry some of the solids into the outlet line; the disposal system following the tank will likewise become flooded or clogged, and may fail. Drainage from garage floors or other sources of oily waste should also be excluded from the tank.

Waste brines from household water softener units have no adverse effect on the action of the grey water treatment tank, but may cause a slight shortening of the life of a disposal field installed in a structured clay type soil.

Adequate venting is obtained through the building plumbing if the tank and the plumbing are designed and installed properly. A separate vent on a grey water treatment tank is not necessary.

A chart showing the location of the grey water treatment tank and disposal system should be placed at a suitable location in dwellings served by such a system. Whether furnished by the builder, grey water treatment tank installer, or the local health department, the charts should contain brief instructions as to the inspection and maintenance required. The charts should assist in acquainting homeowners of the necessary maintenance which grey water treatment tanks require, thus forestalling failures by assuring satisfactory operation. The extension of the manholes or inspection holes of the grey water treatment tank to within 8 inches of the ground surface will simplify maintenance and cleaning.

Abandoned grey water treatment tanks should be filled with earth or rock.

INSPECTION

After a grey water treatment-tank/disposal-field system has been installed and before it is used the entire system should be tested and inspected. The grey water treatment tank should be filled with water and allowed to stand overnight to check for leaks. If leaks occur, they should be repaired. The disposal field should be promptly inspected before it is covered to be sure that it is installed properly. Prompt inspection before backfilling should be required by local regulations, even where approval of plans for the subsurface disposal system has been required before issuance of a building permit. Backfill material should be free of large stones and other deleterious material and should be overfilled a few inches to allow for settling.

Appendix A

Soil Absorption Capacity

GUIDE FOR ESTIMATING SOIL ABSORPTION POTENTIAL

A percolation test is the only known means for obtaining a quantitative appraisal of soil absorption capacity. However, observation and evaluation of soil characteristics provide useful clues to the relative capacity of a soil to absorb liquid. Most suitable and unsuitable soils can be identified without additional testing. When determined and evaluated by trained or experienced soil scientists or soil engineers, soil characteristics may permit further categorizing of suitable soils. This has been done for some areas of the country and described in the soils reports mentioned below.

Soil Maps.—The capacity of a soil to absorb and transmit water is an important problem in agriculture, particularly in relation to irrigation, drainage, and other land management practices. Through studies in these fields, a variety of aids have been developed for judging the absorption of water transmission properties of soils, which could be helpful in the sewage field. Considerable information has been accumulated by agricultural authorities on the relative absorption capacities of specific soils in many areas of the United States. Much of this information is included in Soil Survey Reports and Maps published by the United States Department of Agriculture in cooperation with the various State agricultural colleges. The general suitability of specific soils for effluent disposal may often be interpreted from these reports and maps.

Clues to Absorption Capacity.—Considerable information about relative absorption capacities of soils may also be obtained by a close visual inspection of the soil. The value of such an inspection depends upon some knowledge of the pertinent soil properties. The main properties indicative of absorption capacity are soil texture, structure, color, depth or thickness of permeable strata, and swelling characteristics.

Texture.—Soil texture, the relative proportion of sand, silt, and clay, is the most common clue to water absorption capacity. The size and distribution of particles govern the size and distribution of pores which, in turn, govern the absorption capacity. The larger the soil particles, the larger are the pores and the faster is the rate of absorption.

Texture can best be judged by the feel. The lighter or sandier soils have a gritty feel when rubbed between the thumb and forefinger; silty type soils have a "floury" feel and, when wetted have no cohesion; heavier, clay type soils are dense and hard when dry, and have a slick greasy feel when wetted.

The use of texture as a clue to absorption qualities has its limitations; it is primarily reliable in the sandier soils. In the heavier type soils, including sandy soils containing appreciable amounts of silt or clay, one must look for additional clues, such as structure and soil color, as indicators of absorption capacity.

Structure.—Soil structure is characterized by the aggregation or grouping together of textural particles, forming secondary particles of larger size. Such secondary particles then tend to govern the size and distribution of pores and, in turn, the absorption properties. Structure can easily be recognized by the manner in which a clod or lump breaks apart. If a soil has structure, a clod will break with very little force, along well defined cleavage planes, into uniformly sized and shaped units. If a soil has no structure, a clod will require more force to break apart and will do so along irregular surfaces, with no uniformity in size and shape of particles.

In general, there are four fundamental structure types, named according to the shape of the aggregate particles: platy, prism-like, block-like, and spheroidal. A soil without structure is generally referred to as massive. Spheroidal structure tends to provide the most favorable absorption properties, and platy structure, the least. Although other factors, such as size and stability of aggregates to water, also influence the absorption capacity, recognition of the type of structure is probably sufficient for a general appraisal.

Color.—One of the most important practical clues to water absorption is soil color. Most soils contain some iron compounds. This iron, like iron in a tool or piece of machinery, if alternately exposed to air and to water, oxidizes and takes on a reddish-brown to yellow color. Thus, if a soil has uniform reddish-brown to yellow oxidized color, it indicates that there has been free alternate movement of air and water in and through the soil. Such a soil has desirable absorption characteristics. At the other extreme are soils of a dull gray or mottled coloring, indicating lack of oxidizing conditions or very restricted movement of air and water. These soils have poor absorption characteristics.

Depth or Thickness of Permeable Strata.—The quantity of water that may be absorbed is proportional to the thickness or volume of the absorbent stratum, when all other conditions are alike. In a soil having a foot or more of permeable material above tight clay, absorption capacity is far greater than that of the same kind of material lying within 3 inches of tight clay. When examining soils or studying soil descriptions, the depth and thickness, therefore, are important criteria of absorption capacity.

Swelling Characteristics.—Most, but not all, clays swell upon the addition of moisture. There are many clays (in the tropics, in particular) that do not swell appreciably. There are also some soils in the United States which do not swell noticeably. On the other hand, some soils have a very high percentage of swelling, and these in particular must be suspect. Relative swelling of different soils is indicated by relative shrinkage when dry, as shown by the numbers and sizes of cracks that form. Those that shrink appreciably when dry are soils that may give trouble in a disposal field when they are wet.

Information obtained through inspection or from soil maps and reports can be of particular value in preliminary appraisal of soils for sewage disposal. For instance, in many cases, unsuitable soils may be immediately ruled out on the basis of such information; in other cases, selection of the best of several sites may be made on the basis of the inspection. Absorption capacity information obtained in this manner is relative. For quantitative information upon which to base specific design, we still must depend on some direct measurement, such as a water absorption rate as measured by a percolation test.

EVAPOTRANSPIRATION

In tight clay soils, where absorption is very limited, plant transpiration has been employed with some degree of success in aiding in the disposal of sewage effluent. By placing the distribution lines near the top of the trench, where they are not likely to become clogged with roots, or by laying the lines practically at ground surface and covering with suitable fill material, use may be made of the action of trees, shrubs, and grasses in the absorption and subsequent release to the atmosphere of appreciable quantities of moisture, through the process of transpiration.

Climate is important. Evapotranspiration increases with decreasing humidity and increasing temperature and air turbulence. Length of day and amount of sunshine are further influencing factors. Consequently, evapotranspiration should be of most benefit to disposal fields in the southern portion of the country, where the growing season is longest. In the extreme North, where the growing season extends only from May to September, evapotranspiration would be of minimum value, but may be useful in summer resort areas.

Because of these variations, and equally great variations in the types of plants grown in different parts of the country, as well as the differing transpiration rates in different plants, it would be hazardous to generalize in making specific suggestions on design criteria for systems dependent on evapotranspiration for successful operation.

CURTAIN DRAINS FOR DISPOSAL FIELDS

Even though proper precautions are taken to ascertain the maximum ground water table before a disposal field is approved for a given area, instances occur where the height of a water table rises above that indicated by past performances and observations. This may lead to failure of the disposal system.

It is sometimes possible to rehabilitate such a system by constructing around the disposal area a curtain drain designed to intercept the superficial ground water and carry it away from the area. If the disposal field is on sloping ground, a single drain at the upper end of the area may be sufficient. Where open curtain drains are not feasible, rock-filled trenches above or around the disposal area may sometimes be used. In either case, it is necessary to have a suitable outlet for the ground water that is intercepted.

PERCOLATION TEST HOLES

When making percolation tests in some types of soil, it has been found that the side walls of test holes have a tendency to cave in or slough off, and settle to the bottom. The condition is most likely to occur where the earth is initially dry and overnight soaking is necessary. The caving can be prevented, and more accurate results obtained, by placing in the test hole a wire cylinder surrounded by gravel of the same size that is used in the tile field.

OTHER PERCOLATION TESTS

A Soil Percolation Test for Determining the Capacity of Soils for Absorption Sewage Effluents, by John E. Kiker, Jr.[1]

1. Dig a hole about 1 foot square to the depth at which it is proposed to lay the tile drain.
2. Fill the hole with water and allow the water to seep away. When the water level falls to within 6 or 8 inches of the bottom of the hole, observe the rate at which the water level drops.
3. Continue these observations until the soil is saturated and the water seeps away at a constant rate. (Keep adding water until the rate becomes constant.)
4. Compute the time required for the water to drop 1 inch after the soil becomes saturated. This is the standard percolation time, t.

A Percolation Test for Determining the Capacity of Soils for Absorption Sewage Effluents, by Harvey F. Ludwig, Gordon W. Ludwig, and John Stewart.

While the previously cited percolation tests include procedures generally familiar to workers in this field, the method proposed by Ludwig et al., includes an additional step. The field data are further analyzed graphically to obtain the percolation rate under saturated soil conditions. The mathematical evaluation is based on the premise that the ratio, time in minutes for the water level to drop 1 inch, is constantly increasing during the progress of the test, but at a rate of increase which is constantly decreasing. This hyperbolic relationship is utilized to determine the maximum ratio $\frac{minutes}{inch}$ which is the saturated condition under which water would seep away at a constant rate. A discussion of this method with illustrations and an example is found in the following articles: Improved Soil Percolation Test, Harvey F. Ludwig and Gordon W. Ludwig, *Water and Sewage Works Journal,* vol. 96, no. 5, May 1949, p. 192; Equilibrium Percolation Test for Estimating Soil Leaching Capacity, Harvey F. Ludwig and John Stewart, *Modern Sanitation,* vol. 4, no. 10, Oct. 1952.

---

[1] Reprinted from Subsurface Sewage Disposal, by John E. Kiker, Jr., Bulletin No. 23, Florida Engineering and Industrial Experiment Station, 1948.

Appendix B

Grey Water Drainage Fixture Unit Values

Type of Fixture or Group of Fixtures	Drainage Fixture Unit Value
Automatic clothes washer (2" standpipe)	3
Bathtub (with or without overhead shower)	2
Combination sink and tray with food disposal unit	4
Combination sink and tray with one 1½" trap	2
Combination sink and tray with separate 1½" traps	3
Dental cup or cuspidor	1
Dental lavatory	1
Drinking fountain	½
Dishwasher, domestic (gravity drain)	2
Floor drains with 2" waste	3
Kitchen sink, domestic, with one 1½" waste	2
Kitchen sink, domestic, with food waste grinder	2
Lavatory with 1¼" waste	1
Laundry tray (1 or 2 compartments)	2
Shower stall, domestic	2
Showers (group) per head	2
Sinks:	
Surgeon's	3
Flushing rim (with valve)	6
Service (trap standard)	3
Service (P trap)	2
Pot, scullery, etc.	4
Wash sink (circular or multiple) each set of faucets	2
Unlisted fixture drain or trap size:	
1¼" or less	1
1½"	2
2"	3
2½"	4
3"	5
4"	6

Appendix C

Suggested Specifications for Watertight Concrete

1. *Materials*

Portland cement should be free of hard lumps caused by moisture during storage. Lumps from dry packing that are easily broken in the hand are not objectionable.[1]

Aggregates, such as sand and gravel, should be obtained from sources known to make good concrete. They should be clean and hard. Particle size of sand should range very fine to ¼ inch. Gravel or crushed stone should have particles from ¼ inch to a maximum of 1½ inches in size. Water for mixing should be clean.

2. *Proportioning*

Not more than 6 gallons of total water should be used for each bag of cement. Since sand usually holds a considerable amount of water, not more than 5 gallons of water per bag of cement should be added at the mixer when sand is of average dampness. More mixing water weakens the concrete and makes it less watertight. For average aggregates, the mix proportions shown in the table below will give watertight concrete.

Average Proportions for Watertight Concrete

Max. Size Gravel (in.)	Cement (volume)	Water[1] (volume)	Sand (volume)	Gravel (volume)
1½..............................	1	¾	2¼	3
¾..............................	1	¾	2½	2½

[1] Assuming sand is of average dampness.

3. *Mixing and Placing*

All materials should be mixed long enough so that the concrete has a uniform color. As concrete is deposited in the forms, it should be tamped and spaded to obtain a dense wall. The entire tank should be cast in one continuous operation if possible, to prevent construction joints.

4. *Curing*

After it has set, new concrete should be kept moist for at least seven days to gain strength.

---

[1] Type V portland cement may be used when high sulfate resistance is required.

Bibliography

Partial List of References on Individual Sewage Disposal Systems

Appendices to Study of Seepage Beds, James B. Coulter; Thomas W. Bendixen; and Allen B. Edwards; Report to the Federal Housing Administration, December 15, 1960.

A Basis for Classifying Soil Permeabilities, T.W. Bendixen; M.F. Hershberger; and C.S. Slater; Jour. Agric. Research, v. 77, no. 5; Sept. 1, 1948.

Causes and Prevention of Failure of Septic Tank Percolation Systems, P.H. McGauhey; and John H. Winneberger; Federal Housing Administration Report No. 533, 1964.

Degradation of ABS and Other Organics in Unsaturated Soils, G.G. Robeck; J.M. Cohen; W.J. Sayers; and R.L. Woodward; Journal of the Water Pollution Control Federation; v. 35; no. 10; Oct. 1963.

Detergents and Septic Tanks, James E. Fuller; Sewage and Ind. Waste, 24, 844; July 1952.

The Effect of Automatic Sequence Clothes Washing Machines on Individual Sewage Disposal Systems, NAS Publication 442, Building Research Institute, National Academy of Science, National Research Council, Washington, D.C.

Effects of Food Waste Grinders on Septic Tank Systems, T.W. Bendixen; A.A. McMahan; J.B. Coulter; and R.E. Thomas; Report to the Federal Housing Administration, Nov. 15, 1961.

Effects of Ground Garbage on Sewage Treatment Processes, William Rudolfs; Sewage Works Journal, v. 18, no. 6, p. 1144; Nov. 1946.

Effectiveness of the Distribution Box, J.B. Coulter; and T.W. Bendixen; Report to the Federal Housing Administration, Feb. 19, 1958.

Environmental Sanitation, J.A. Salvato, Jr.; John Wiley & Sons, Inc. 1958.

Estimating Soil Moisture Conditions and Time for Irrigation with the Evapotranspiration Method, C.H.M. Van Bavel; U.S. Department of Agriculture Publication ARS 41-11, August 1956.

Factors Influencing the Design and Operation of Soil Systems for Waste Treatment, G.G. Robeck; T.W. Bendixen; W.A. Schwartz; and R.L. Woodward; Journal of the Water Pollution Control Federation, v. 36, no. 8, Aug. 1964.

Feasibility Study of Holding Tanks and Sewage Haulage System for Individual Premises, J.V. Morris; L.N. Hogarth; and W.L.C. Knowles; Ministry of the Environment, Ontario, Canada, A Report Prepared by James F. MacLaren Limited, March 1973.

Field Investigation of Waste Water Reclamation in Relation to Ground Water Pollution, State Water Pollution Control Board, Publication No. 6, State of Calif., 1953.

Final Report on a Study of Preventing Failure of Septic Tank Percolation Systems, P.H. McGauhey and J.H. Winneberger; SERL Report No. 65-17, Sanitary Engineering Research Laboratory, University of California, Berkeley, 1965.

Improved Soil Percolation Test, H.F. Ludwig and G.W. Ludwig, Water and Sewage Works, v. 96, 5; May 1949.

Individual Sewage Disposal Systems, Recommendations of Joint Committee on Rural Sanitation; Reprint No. 2461 from Public Health Reports, Public Health Service (revised 1947).

Manual of Septic-Tank Practice, U.S. Public Health Service, PHS Publication No. 526, Reprinted 1969.

Notes on the Design of Sewage Disposal Works, with Special Reference to Small Installations, Henry Ryon; Albany, N.Y.; 1928.

Progress in the Design of Rural Sewage Disposal Systems, Am. Jour. Pub. Health, Year Book, Part 2, 1952-53; May 1953.

Regulations and Information Concerning Sanitation and Single Family Homes and Small Communities, National Association of Public Health Enterprises, Paris 8e, 3d. ed., 1969.

Septic Tank Care, PHS Publication No. 73, U.S. Department of Health, Education, and Welfare, Public Health Service, Washington, D.C.

Septic Tank Drainage Systems, W.H. Sheldon; Research Report No. 10, Michigan State University, Agriculture Experiment Station, East Lansing.

Septic Tank Soil Absorption Systems for Dwellings, Construction Aid No. 5, Housing and Home Finance Agency, Division of Housing Research, Washington 25, D.C.

Soils Suitable for Septic Tank Filter Fields, W.H. Bender; U.S. Department of Agriculture, Agricultural Information Bulletin No. 243, 1961.

Soil Survey Manual, Agricultural Handbook No. 18, U.S. Department of Agriculture, 1951.

Some Factors Which Modify the Rate and Total Amount of Infiltration of Field Soils, G.W. Musgrave; and G.R. Free; Jour. Amer. Soc. Agron., v. 28; 727-739; 1936.

Studies on Household Sewage Disposal Systems, Part I, S.R. Weibel; C.P. Straub; and J.R. Thomas; Federal Security Agency, Public Health Service, Environmental Health Center, 1949.

Studies on Houshold Sewage Disposal Systems, Part II, T.W. Bendixen; M. Berk; J.P. Sheehey; and S.R. Weibel; Federal Security Agency, Public Health Service, Environmental Health Center, 1950.

Studies on Household Sewage Disposal Systems, Part III, S.R. Weibel; T.W. Bendixen; J.B. Coulter; U.S. Dept. of Health, Education, and Welfare, Public Health Service, Robert A. Taft Sanitary Engineering Center, 1955.

A Study of Methods of Preventing Failure of Septic Tank Percolation Systems, J.H. Winneberger and P.H. McGauhey, SERL Report No. 65-16, Sanitary Engineering Research Laboratory, University of California, Berkeley, 1965.

Study of Seepage Beds, J.B. Coulter; T.W. Bendixen; and A.B. Edwards; Report to the Federal Housing Administration, Dec. 15, 1960.

Study of Seepage Pits, Thomas W. Bendixen; R.E. Thomas; and J.B. Coulter; Report to the Federal Housing Administration, May 1, 1963.

A Study of Serial Distribution for Soil Absorption Systems, G.M. Sullivan; J.B. Coulter; and T.W. Bendixen; Report to the Federal Housing Administration, April 8, 1959.

Subsurface Sewage Disposal, J.E. Kiker, Jr.; Bull. No. 23, Florida Eng. Ind. Exper. Station, Univ. of Florida, Dec. 1948.

Transpiration and Total Evaporation, Physics of the Earth, IX, Charles H. Lee, Hydrology; McGraw-Hill Book Co., Inc.; 1942.

Underdrained Filter Systems, Whitby Experimental Station, N.A. Chowdhry; Ministry of the Environment, Ontario, Canada, Interim Report-Part 1, 1972, and Interim Report-Part 2, 1973.

Water, The Yearbook of Agriculture, 1955, U.S. Dept. of Agriculture, 84th Cong., 1st Session, House Document No. 32.

Part II

Characterization of Grey Water and Soil Mantle Purification

SEPARATION OF WATER-BORNE WASTES

J.T. Winneberger

Water pollution has been defined as the "inability of man to recycle wastes for the benefit of all living creatures."[1] When making that remark, Warshall was considering the legitimate place in the biosphere where man's items of disclaimed ownership belonged. Western man has come to regard himself as separate of the remainder of living things, and indeed he has mostly troublesome vestiges of adaptive, instinctive behavior he once had. Only in recent years is he beginning to notice a need to live in harmony with his environment. When scientific attitudes replace prejudices and man develops a genuine respect for all living things, it will occur to him that his wastes must be offered in palatable form to legitimate recipients, be they animal, vegetable, or mineral.

WORSHIP OF BIGNESS

Scientists acquainted with environmental studies termed *ecology* have been painfully aware that man's worship of bigness has brought about ecological disasters, which at their inception were commonly believed to be worthwhile projects. Indeed, creation of dams, highways, megalopolises, and hydrogen bombs testify to spectacular engineering skills. Still, ecologists question the wisdom of committing many of such grand projects at all.

Man's worship of bigness caused him to readjust continental plumbing to compensate for overlooked needs of the megalopolises he created.[2] At the same time he found it convenient to collect wasted waters and conduct them in increasing volumes to as few localities as possible, engineers admonished, "dilution is the solution to pollution." They failed to notice that conduction of great collected volumes to a single place was contrary to the concept of dilution. Collection amounted to concentration.

NEW DIRECTIONS

After a history of facing complexities of problems created by collecting sewages together, Mr. Charles Carnahan spoke to the need of reversing engineering thinking by considering management of wastewaters nearer their sources.[3] Environmentalists had already begun to question the wisdom of massive sewer projects. Still others had begun to question the wisdom of national support of sewering programs, which could not be afforded by the

John Timothy Winneberger, Ph.D., Consultant, Individual Sanitation Systems, 1018 Hearst Avenue, Berkeley, California 94710.

various smaller communities comprising a meaningfully large part of the national community.

Whatever the reasons may be, progressive architects and engineers are beginning to seek tools for managing wastewaters in smaller volumes. It has also become of interest to reduce volumes of waters used to transport wastes, and to keep wastes separated in efforts to simplify their respective treatments.

BENEFITS OF FLOW REDUCTION

Many thoughts come to mind when considering reducing volumes of waters used to transport wastes, among which are: lot sizes are frequently limited by older and hazardous on-site types of disposal systems; treating fresh waters to drinking qualities and then using almost half to only flush household toilets constitutes incredible waste; and, fresh water supplies in most places are finite and, as such, are not unlimited in availability for wasteful uses. Those thoughts point to benefits of reducing flows. Another benefit of flow reduction is elaborated below, as an example.

In the United States, the sewage disposal needs of about one in four all-year dwelling units are served by a septic-tank or cesspool (terminology reflecting confusion of subsurface disposal with treatment devices). Altogether, there must be 16 million to 17 million subsurface disposal systems constituting a national investment of perhaps three billion dollars. It is known that subsurface disposal systems have a lifespan which is related to the wastewater and volumes of sewage concentrations disposed to them; sometimes, but not always, are relationships in direct proportion. For example, a kind of subsurface disposal system (often termed, "seepage bed") constructed in Decatur and Sequoia soils in Tennessee may have a lifespan of five years if the system is 400 square feet in horizontal size. But, if it is twice as large (800 sq.ft.), its lifespan is not just twice as long, rather, it can be expected to be 20 years–four times longer.[4]

Inasmuch as disposal systems are constructed in direct proportion to expected wastewater flows, the value of reducing both flows and sewage strength becomes obvious–lifespans should be much lengthened. Considering this with the national investment leads to the inescapable conclusion that such considerations are unbelievably worthwhile, at the least in gaining more from existing investments.

GREY WATER

When considering reduction of wastewater flows, the biggest water-wasting device that first comes to mind is the water flush toilet. Almost everyone has participated in the use of about five gallons of drinking-quality water to flush out of sight, objects of no more consequence than a cigarette butt. Altogether, waters wasted in flushing toilets approach almost half of all water used in a household.

With the recent advent of aesthetically acceptable toilets which do not use water at all, there was a need for a term to refer to wastewaters emanating from other household devices. These wastewaters have been collectively termed *grey water,* and being human, we have expressed our prejudices against our body wastes by terming them *black.* The day is coming when western man will recognize "black wastes" to be what they really are: legitimate parts of natural cycles. They could be used to man's advantage, and perhaps soon they will be regarded as a valuable resource. Meanwhile, they remain "black" and, in efforts to rid ourselves of those unwanted materials without fouling immediate environs, it seems easier to handle black wastes separately of grey.

A. *Volumes*

It is common information to persons acquainted with subsurface disposal systems (sanitarians, contractors, pumpers, etc.), that homeowners with on-site systems use meaningfully less water than do homeowners connected to public sewerage. It is also known that water costs (wells vs. public supplies, metered vs. unmetered waters, etc.) have profound effect on usage. Consequently, when considering "average flows," authorities vary greatly in opinions. Still, in the usual absence of local knowledge, there is need for estimates, however crude but hopefully accurate. In surveying literature, one group of investigators reported of the household: 45% of water usage for toilet flushing; 30% for bathing; 6% for the kitchen; 5% for drinking; 4% for the laundry; 3% for cleaning; 3% for sprinkling; 1% for auto washing; and, 3% used for other miscellaneous purposes.[5] As would be expected, proportions of wastewater flows from different household uses vary greatly with total water usage. Water used for toilet flushing varies almost twofold, depending on which day of the week it is. After discussing extreme vagaries in wastewater volumes, investigators provided a description of expected flows from an "average household of two adults and two children." The household was assigned 255 gallons per day (gpd) of potable waters, which emanated as wastewaters: 27 gpd from the kitchen; five gpd from the utility sink; 35 gpd from laundry; 80 gpd from bathing; eight gpd from lavatory use; and, 100 gpd from toilet use. By assuming a waterless toilet, the above data can be used to arrive at the proportions of grey water emanating from such an "average household." We have: about 17% from the kitchen; 3% utility sink; 23% laundry; 52% bathing; 5% lavatory; and, of course, none from the waterless toilet.

B. *Quality*

Studies in Sweden reported data reflecting what is common information among technicians, but not commonly admitted: technicians are not equally competent.[6] Consequently, replicated samples in different laboratories were reported with BOD values in variance 10% to 30%. Thus, analyses can provide meaningful sources of errors. Along with analytical data, the Swedish studies reported of volume and nature of flows:

1. "variations within the day itself were so great and so unsystematic that the analysis results from separate parts of the day do not supply any usable representative figures of the quantities of pollution";

2. "If the composition of grey water is first studied, it is found that nearly 70% of the pollution in terms of BOD_5 emanates from the kitchens. Of the quantities of phosphorus, nearly 60% originates from the laundry,

and only 14% from the kitchens, while the quantities of nitrogen are split up in such a way that kitchens supply 51%, the bathrooms 31%, and the laundry 18%."

3. In comparing pollutants of grey and black waters, "the share of grey water is higher for BOD_5, phosphorus, dry matter, fixed and volatile total residues, as well as plate count 35° and Coli 35°. Grey water contributes between 53 and 70% in these cases."

4. "Particularly noticeable is the part played by black water as the principal source of nitrogen (91% of the total amount of nitrogen)," and

5. in studying susceptibility of BOD of wastes to treatment, it was found that, "black water from toilets decomposes relatively slowly (compared to grey), i.e., approximately at the same rate as municipal waste water."

Clearly, grey waters are variable in qualities, but susceptible to treatment. This suggests that anaerobic processes, known to be more stable than aerobic processes to shock loadings,[7] should offer acceptable treatment of grey water. Such a technique, in fact, is recommended for everyday treatment of grey water where soil conditions are suitable. A conventional anaerobic septic tank/ leach field adapted for use with grey water only is a feasible system of handling on-site wastewater treatment and disposal. Knowing that the size of the system is governed by both the soil percolation rate and by the flow, it is not unreasonable to consider a similar sizing relationship for waste strength. Rein Laak, who contributed one of the following articles, conclusively determined that the clogging rate of a leach field and thus its area is dependent upon both the BOD_5 and the total suspended solids load in this relationship:[8]

$$\text{ADJUSTED AREA REQUIRED} = \left(\begin{matrix}\text{AREA REQUIRED FOR}\\ \text{STANDARD SEPTIC TANK}\\ \text{PRETREATMENT}\end{matrix}\right)\sqrt[3]{\frac{BOD_5 + TSS}{250}}$$

This knowledge will enable the designer of a sewerless home disposal system to scale the treatment system to the known strength and flow of grey water waste and assure himself that for a particular class of soil, the system has every reasonable expectancy of proper performance and longevity.

Thus, while ultimate disposal of grey water might one day include reuses such as garden watering, car washing, or perhaps in helping with household chores, in the immediate future, despite Winneberger's Law II*, we look forward to increasing interest in smaller versions of customary subsurface disposal fields for handling grey water.

---

*Winneberger's Law II states, "Interests of authorities in wastewater discharges are inversely proportional to volumes." Whereas discharge of 10% of municipal BOD into water courses would be considered a worthwhile achievement (90% reduction in a secondary plant of exceptional quality), acceptability of on-site disposal systems includes functioning perfectly and never discharging directly to water courses, even of minor consequence.

THE FOLLOWING CHAPTERS

The following chapters are selected sources of data to be used by persons interested in the nature of grey water. Very useful information has been included on subsurface disposal technology, inasmuch as such techniques are commonly used and not widely understood.

While considering data collected at one place or another, it should be thoroughly borne in mind that grey water would vary from time-to-time, place-to-place, and in almost every conceivable way. Regardless of how statistically meaningful any one study would be, the data would not precisely describe all other situations. Nevertheless, the data herein were intended to be contributions to a growing technology. In absence of local knowledge, they should prove to be invaluable tools for local applications since nothing comparably descriptive exists elsewhere.

REFERENCES

[1]Warshall, P. and J.T. Winneberger. *Septic-Tank Practices.* Bolinas, Calif.: Shelter Publications, Box 279, Bolinas, Calif. 94924. 1974.

[2]Nace, R.L. "Arrogance towards the Landscape." *Water Spectrum.* Washington, D.C.: Dept. of the Army, Office of the Chief of Engineers. Winter 1970.

[3]Mr. Charles Carnahan, Retired Executive Officer, California Regional Water Quality Control Board, Central Valley Region. In an address to Water Pollution Control Engineers Association. 12 October 1971.

[4]Coulter, J.B., T.W. Bendixen, and A.B. Edwards. *Study of Seepage Beds–Part I, Knox County, Tennessee.* A report to the Federal Housing Administration, Cincinnati, Ohio: USPHS, Robt. A. Taft San. Eng. Center. 14 February 1960.

[5]Bailey, J.R., R.J. Benoit, J.L. Dodson, J.M. Robb, and H. Wallman. *A Study of Flow Reduction and Treatment of Waste Water from Households.* A report for the Federal Water Quality Administration of the Department of the Interior, Program #11050FKE, Contract #14-12-428, FWQA Project Officer, C.L. Swanson, Advanced Waste Treatment Research Laboratory, Cincinnati, Ohio. Groton, Conn.: General Dynamics, Electric Boat Division. December 1969.

[6]Olsson, E., L. Karlgren, and V. Tullander. "Household Waste Water." The National Swedish Institute for Building Research, Report 24. 1968.

[7]Winneberger, J.H., L. Francis, S.A. Klein, and P.H. McGauhey. *Biological Aspects of Failure of Septic-Tank Percolation Systems–Final Report.* Berkeley, Calif.: San. Eng. Res. Lab., Univ. of Calif. 31 August 1960.

[8]Laak, R., *The Effect of Aerobic and Anaerobic Household Sewage Pretreatment on Seepage Beds,* Doctoral Thesis, University of Toronto, 1966.

RELATIVE POLLUTION STRENGTHS OF UNDILUTED WASTE MATERIALS DISCHARGED IN HOUSEHOLDS AND THE DILUTION WATERS USED FOR EACH

Rein Laak

INTRODUCTION

By keeping records and weighing the quantities of consumer products used in five households and by measuring the water used for each plumbing fixture, a detailed pollutant budget was obtained. Laboratory analysis for consumer products used and actual samples of the wastewater for BOD, COD, ammonia, Kjeldahl and nitrate nitrogen, and inorganic phosphates are presented in table form for design of recycle systems, estimating pollution loads and calculating the impact of consumer brand changes. Consumer products contributed less than 50% of the total pollution load. About half of the waste flow, 90% of the nitrogen, 60% of COD, and 50% of the phosphates were contributed by toilet wastes. Effective environmental protective measures appear to be reduction or elimination of water-carried toilet systems and partial recycling of water between bathing and laundry.

PROJECT OBJECTIVES

To determine the pollutional load and measure the dilution water used by each type of household plumbing fixture. To establish a method of estimating the pollution loads and to provide basic data for further studies concerning plumbing fixture redesigns and methods of transporting water and wastewater.

ACHIEVEMENT OF OBJECTIVES

All objectives were achieved and data put into tabular form in Tables 1-9; the following discussion gives these achievements in detail.

RESEARCH PROCEDURES USED

In order to collect the data, five families were chosen for investigation. These families were designated A,B,C,D, and E respectively. The number of people, house type, plumbing fixtures' number or capacity are listed in Table

Rein Laak, Ph.D., P.E., Associate Professor of Civil Engineering, School of Engineering, University of Connecticut, Storrs, Connecticut 06268.

1. Fragmented data were collected from three additional families; however, the data, although comparable, were not included in this report.

For each family the following were carried out:

Quantities of Water Used for Each Plumbing Fixture (See Table 2)

A. In a kitchen sink, bathroom sink, and a bathtub, measured volumes of water were poured into each basin, and the relation between depth and capacity of each sink or tub was obtained. Measurements of the water depth before draining the wastewater from each plumbing fixture were recorded by each family.

B. For laundry washing machines quantities of water used for each laundry load per day were used to calculate the water consumption.

C. For toilet flush water, the volume of water per flush was calculated from tank dimensions. Counters activated by the flush lever were installed to record the number of flushes. In total, over 10,000 flushes were recorded with an overall mean of about five flushes per person per day. The quantity of water flushed per capita per day for each family was calculated from the actual number of flushes.

D. For some families the kitchen sink, bathroom sink, or bathtub drain plug leaked. Here the quantities of water used were estimated by measuring the flow rate and the duration of flow.

Wastewater Sampling Analyses (See Table 3)

The wastewater was sampled immediately after dishwashing, bathing, laundering, hand or face washing, and tooth brushing. The samples were analyzed in the laboratory for biochemical oxygen demand (BOD_5), chemical oxygen demand (COD), nitrogen compounds (NO_3-N, NH_3-N), and inorganic phosphate (PO_4). By these parameters the pollution strength of the wastewater was defined.

In addition to these analyses, wastewaters from kitchen sinks for families A,B, and E were also analyzed for total Kjeldahl nitrogen (TKN). The results of the analyses showed a range of 4 - 30 mg/1 (TKN). The water closet wastes showed a calculated concentration of 193 mg/1 (see Table 4 for measured TKN values for feces and urine).

Feces and Urine Analyses (See Table 4)

Wastewater from water closets was not sampled. Instead, fresh feces from two persons and urine from student washrooms were collected on 12 separate occasions and analyzed. It should be pointed out that the samples analyzed were prepared by dissolving 0.1 gram of wet feces or one milliliter of urine in one liter of distilled water. The average discharge of urine was estimated at 1,200 ml/day per adult and 800 ml/day per child; wet fecal output was estimated at 130 g/day per adult and 90 g/day per child. The weighed average feces and urine concentration in the toilet wastewater was calculated from this (Table 4A).

Manufactured Materials (See Table 5)

The significant type of materials or consumer products used at each plumbing fixture were:

kitchen sink:	dishwashing liquid, cleanser, cleaner ammonia, soap
bathtub:	soap, shampoo
bathroom sink:	soap, toothpaste, mouthwash
laundry machine:	detergent, bleach, softener
water closet:	toilet tissue paper (no deodorants were used)

Families A and E were measured over six months by recording the period over which the contents of packages were consumed.

Families B,C, and D were measured by weighing and recording once a week, over a two-month period, the brands of materials used.

In order to measure the pollution load contributed by the manufactured materials, the brands of materials used in each family were independently purchased and analyzed. Excluded in this analysis were the pollution from leftovers discharged and from soil and sweat on clothes and body.

Table 5 summarizes the data and shows the mean pollution strengths of the products in terms of mean daily usage. Table 6 combines the human excrement and manufactured materials. Table 7 expresses Table 6 data in terms of concentrations. Table 8 shows Table 3 data in terms of mg/c/d and adds the pollution from the water closet.

Analytical Work

The biochemical oxygen demand, chemical oxygen demand, and total Kjeldahl nitrogen of wastewater in this study were determined by the procedures in *Standard Methods for the Examination of Water and Wastewater,* 12th Edition. Other nitrogen compounds and inorganic phosphate were determined by using a Hach Co. colorimeter and procedures.

It could be that the mean and the range of values found for the five households investigated may not show the true mean and the variation of values for all of the households in the U.S.A. However, the study showed that a large fluctuation in the quantity and quality of waste materials and diluting water occurs between households and between plumbing fixtures. Data in Table 2, showing total water use of 26 to 65 gal/cap/day, fit the reported data in literature.[9] The water use for each plumbing fixture also falls within the range of the values reported in the literature (see Table 9).

The values obtained for the manufactured materials wasted and the excrement appeared to agree closely with some of the values reported. For example: phosphate (as inorganic) excreted per day varies from 0.34g to 2.2g per 24 hours with urine N:P ratio of 10:1.[5] It is reported that the phosphate discharge in urine is highest during the afternoon with a six-fold hourly variation.[11] The phosphate values measured range from 4.2g to 7.7g/24 hours (samples taken during an eight-hour work day). The ratio of means of measured N:P was calculated to be approximately 2.3:1. The high phosphate was probably due to samples gathered only during the daylight hours. A ratio of 4:1 for N:P has also been reported.[3] A reported BOD_5 of 12,000 mg/1[3] in urine closely agrees with mean range of 5,500 to 13,000 mg/1 found here. Total nitrogen of approximately 11,000 mg/1[8] and 13,000 mg/1[3] in urine was within TKN values measured 10,600 to 15,300 mg/1. Fecal analysis showed that TKN of 0.5g to 2.7g per person per day and phosphate 0.26g to 1.6g per person per day is within the reported values of 1.3g[3] and 1.5g to 1.8g[8] of total nitrogen and 0.5g[3] and 0.3g to 0.6g[8] of phosphate.

The total excrement per capita per day was reported to be 11g of TKN 1.6g phosphate and 20g BOD_5.[8] Previous work reported in reference[8] showed 12g to 14.4g of TKN, 0.9g to 2.6g phosphate and 60g to 100g BOD_5 per capita per day. The measured mean values were 14g of TKN, 6g of phosphate and 20g BOD_5. Comparing the BOD_5 per capita per day values reported for kitchen 20.4g, laundry 2.6g and bathroom 4.6g[8] to measured mean values (Table 8) for kitchen 9.2g, laundry 7.9g and bathroom 8g, it shows that a wide variation is possible. However, the reported values[8] were within the range of values measured.

For the basic data the range of values were shown with arithmetic mean and weighed average. For combinations of data, mean values were used for quantity and concentration calculations. The most relevant data were summarized in Tables 2, 5 and 8. Change in Table 7 total values can be visualized when 25% of households would use low phosphate detergents. The soap detergents would significantly increase the BOD and COD and slightly decrease PO_4. Tables 6 and 8 offer the opportunity to compare the two different measurements in the study. The higher values shown with direct sampling were expected, because the material measurements did not include the food leftovers, grease, etc. . . in the kitchen sink; the body dirt in the bathtub water; and the laundry dirt in the laundry water. Other factors also influenced the differences in values, for example: the use of mean values for calculation, and percent of certain brands used and their pollution parameter concentrations. The latter differences in values would have decreased if a larger number of homes could have been surveyed.

Several observations can be made. The least polluted water was from the bathtub and from the hand washbasin. The toilet contributed (see Table 8) about one-half of the BOD, about one-half of the phosphates, and the majority of nitrogen compounds from a household. Looking at Table 5 and Table 6 it is clear that the resistance to biodegradation (using the ratio of COD to BOD as the criteria for comparison) and the largest bulk of solids were contributed by toilet tissue paper and feces. It would appear that the toilet wastes would be the most difficult to treat. Another observation that can be made is that the consumer products contributed (Tables 5 and 8) less than one-third of the total BOD, about 50%. of PO_4 and an insignificant amount of nitrogen into the wastewater from the households.

CONCLUSIONS

Consumer products wasted in households surveyed contributed less than 50% of the total pollution load in the wastewaters. The waste constituents most resistant to biodegradation were the toilet wastes which contributed about half of the waste flow, approximately 90% of the nitrogen, over 60% of COD and about 50% of phosphates.

Waste flow, consumer products, and excrement quantities measured showed a several-fold variation between households. The type of treatment process proposed would indicate the adjustment necessary to the calculated mean values.

For individual houses, the quantity of three materials used (soap, detergent and toilet paper) and the number of occupants would give sufficient data to estimate the total BOD load, nitrogen (90% from urine), and the amount of phosphates (50% from detergents, 50% from urine) discharged.

The direction to take for household wastewater disposal for reducing potential water pollution and health hazards is to eliminate or reduce water used for toilets. The second step is to recycle water between bathing and laundry. Total recycling of all household wastes, especially toilet wastes, does not appear to be economical, rather a partial recycle of compatible liquids seems feasible.

REFERENCES

[1] Standard Methods for the Examination of Water and Wastewater, 12th Edition, APHA, AWWA, WPCF, 1965.

[2] Bailey, J.R., et al., 1969. "A Study of Flow Reduction and Treatment of Waste Water from Households," FWQA, Dept. of the Interior, WPCRS 11050 FKE 12/69.

[3] Masselli, J.W. and Masselli, N.W., 1970. "Controlling the Effects of Industrial Wastes and Sewage Treatment." New England Interstate Water Pollution Control Commission. TR-15.

[4] Watson, K.S., Farrell, R.P. and Anderson, J.S., 1967. "The Contribution from the Individual House to the Sewer Systems." J.W.P.C.F., *39*, No. 12, p. 2039.

[5] Communication with Windham Community Memorial Hospital. Mr. E. Recor, Chemist, Willimantic, Connecticut.

[6] Laak, R., 1966. "The Effect of Aerobic and Anaerobic Household Sewage Pretreatment on Seepage Beds." Ph.D. Thesis, University of Toronto, Canada.

[7] Dufor, C.N. and Becker, E. "Public Water Supplies of the 100 Largest Cities in the United States." U.S. Geol. Survey, Water Supply Paper 1812, 5 (1964).

[8] Olsson, E., Karlgren, L. and Tullander, V. "Household Waste Water." The National Swedish Inst. for Building Research. Report 24:1968.

[9] U.S. Dept. of Housing and Urban Development, 1961-66. "A Study of Residential Water Use." F.H.A., Washington, D.C.

[10] Hubbell, J.W., 1962. "Commercial and Institutional Wastewater Loadings." J.W.P.C.F. *34*, No. 9, pp. 962-968.

[11] Henry, R., 1965. "Principles of Clinical Chemistry." Hoeter Publ.

ACKNOWLEDGEMENT

The work upon which this publication is based was supported in part by funds provided by the United States Department of the Interior as authorized under the Water Resources Research Act of 1964, Public Law 88-379.

The support of the Institute of Water Resources is gratefully acknowledged. Graduate Students Mr. James Huang and Mr. An-I Lin carried out the field work and the laboratory analysis. The housewives of the five families studied displayed infinite patience.

Table 1. The Five Families

Family	Number of People		House or Apartment	Plumbing Fixtures' Number or Capacity				
	Adults	Children		Kitchen Sink	Bathtub	Bathroom Sink	Laundry Machine	Water Closet
A	2	2(7&9)*	2 Bdrm. Apt.	1	1	1	40 Gal. Load	5.4 Gal. Flush
B	2	0	2 Bdrm. Apt.	1	1	1	35 Gal. Load	4 Gal. Flush
C	2	2(2&5)	2 Bdrm. Apt.	1	1	1	L = 32[t] M = 25 S = 18 Load	4 Gal. Flush
D	2	2(3&5)	3 Bdrm. House	1	1	1	20 Gal. Load	4 Gal. Flush
E	2	0	2 Bdrm. Apt.	1	1 Shower	1	15 Gal. Load	2.2 Gal. Flush

*The numbers in the parentheses indicate the ages of the children.

[t]L = Large Load, M = Medium Load, S = Small Load

Mean number of persons per family = 3.2

Table 2. Mean Water Consumption Rates from Each Plumbing Device (gal/capita/day)

Family Name	Kitchen Sink GPCD	%	Bathtub GPCD	%	Bathroom Sink GPCD	%	Laundry Machine GPCD	%	Water Closet GPCD	%	Total GPCD	%
A	3.2	4.9	15.4	23.5	3.0	4.6	14.3	21.9	29.5	45.0	65.4	100
B	9.1	16.3	5.3	9.4	3.2	5.7	2.1	3.7	36.4	65.0	56.1	100
C	3.4	13.0	5.9	22.3	1.5	5.6	4.3	16.3	11.2	42.8	26.3	100
D	2.1	7.1	5.0	16.8	1.0	3.4	7.9	26.6	13.7	46.2	29.7	100
E	2.1	6.5	10.0	30.9	2.7	8.2	4.5	13.8	13.2	40.6	32.5	100
Average (Weighted)	3.6	9.0	8.5	20.7	2.1	5.1	7.4	18.4	19.8	46.7	41.4	100
Range	2.1- 9.1	4.9 16.3	5.0- 15.4	9.4 30.9	1.0- 3.2	3.4 8.2	2.1- 14.3	3.7 26.6	11.2- 36.4	40.6 65.0	26.3- 65.4	100

1 gal/capita/day = 3.785 liters/capita/day

Table 3. Pollution Concentration of Wastewater as Sampled from Each Plumbing Fixture (mg/1) Mean Values

	BOD	COD	NO_3-N	NH_3-N	PO_4
Kitchen Sink	676	1,380	.56	5.44	12.7
Bathtub	192	282	.36	1.34	.94
Bathroom Sink	236	383	.28	1.15	48.8
Laundry Machine	282	725	1.26	11.3	171.0

Table 4. 0.1g of Feces/ℓ or 1 ml of Urine/ℓ Analysis, Pollution Strength Concentrations mg/1

Pollutant	BOD Mean	BOD Range	COD Mean	COD Range	NO_3-N Mean	NO_3-N Range	NH_3-N Mean	NH_3-N Range	PO_4 Mean	PO_4 Range	T.K.N. Mean	T.K.N. Range
Feces	9.6	5.6 15.9	28.7	16.8 48.2	.003	.002 .005	.15	.09 .22	.58	.23 1.39	1.12	.47 2.36
Urine	8.6	5.3 13.0	17.5	11.6 28.5	.012	.000 .028	2.49	1.5 5.0	5.50	4.0 7.4	12.6	10.1 14.6

Table 4A. Approximate Daily Feces and Urine Production Rate Capita/Day

Pollutant	1 Adult	1 Child	Weighted Average For 10 Adults and 6 Children
Feces	130 g	10 g	115 g
Urine	1,200 ml	80 ml	1,050 ml

Table 5. Pollution Strength of Manufactured Materials (mg/c/d)

	BOD		COD		NO_3-N		NH_3-N		PO_4	
	Mean	%	Mean	%	Mean	%	Mean	%	Mean	%
Kitchen Sink	2,380	17.1	6,700	18.4	5.5	22.4	58.7	83.5	271	6.8
Bathtub	3,870	27.8	5,570	15.3	3.9	15.8	1.3	1.8	26	.6
Bathroom Sink	2,940	21.1	3,810	10.5	.4	1.6	2.1	3.0	455	11.3
Laundry Machine	1,260	9.1	3,440	9.5	14.5	58.9	8.2	11.6	3,262	81.3
Water Closet (Tissue Paper)	3,450	24.8	16,600	45.5	.0	.0	.0	.0	27	.7
Total	13,900	100.0	36,390	100.0	24.6	100.0	70.3	100.0	4,013	100.0

Table 6. Pollution Strengths of Human Excrement and Manufactured Materials (mg/c/d)

	BOD		COD		NO_3-N		NH_3-N		PO_4		T.K.N.	
	Mean	Range	Mean	Range	Mean	Range	Mean	Range	Mean	Range	Mean	Range
Feces	11,050	6,440-18,300	33,000	19,300-55,400	3.45	2.30-5.75	172	104-253	666	264-1,600	1,290	540-2,710
Urine	9,040	5,560-13,650	18,380	12,200-30,000	12.6	0.-29.4	2,610	1,580-5,250	5,780	4,200-7,770	13,200	10,600-15,300
Total Human Waste	20,090	12,500-31,950	51,380	22,500-85,400	16.05	2.30-35.15	2,782	1,684-5,503	6,446	4,464-9,370	14,490	11,140-18,010
(Table 5) Total Materials	13,900	2,260-43,900	36,390	10,800-95,700	24.6	.4-103.5	70.3	.3-257.8	4,013	583-12,350	—	—
Total Pollution	33,990	14,260-75,850	87,770	33,350-181,100	40.65	2.34-138.65	2,852.3	1,684.3-2,760.8	10,459	5,047-22,720	—	—

Table 7. Pollution Concentrations (mg/1)

		BOD	COD	NO_3-N	NH_3-N	PO_4	T.K.N.
Materials Consumed Only	Kitchen Sink	175	492	.405	4.22	20.0	—
	Bathtub	120	173	.117	.04	.8	—
	Bathroom Sink	372	482	.040	.27	57.6	—
	*Laundry Machine	45	133	.528	.29	116.5	—
	Water Closet Tissue Paper	46	221	.000	.00	.4	—
	Total Materials	89	232	.157	.45	25.6	—
Human Waste	Water Closet Feces and Urine	267	675	.214	37.1	77.0	193
	Water Closet Total	313	896	.214	37.1	77.4	—
	Total Pollution	216	570	.255	18.2	66.6	—

*In these calculations only high phosphate detergents were considered. Using low phosphate soap detergents, values will change.

NOTE: Table 7 was obtained by using Tables 5 and 6 and dividing by values from Table 2.

Table 8. Pollution Loads of Wastewater as Sampled from Each Plumbing Fixture (mg/c/d)

	BOD		COD		NO_3-N		NH_3-N		PO_4		T.K.N.
	Mean	%	Mean	%	Mean	%	Mean	%	Mean	%	Mean
*Kitchen Sink	9,200	19	18,800	16	7.6	10.4	74	2.3	173	1.5	—
*Bathtub	6,180	13	9,080	8	11.6	16.0	43	1.3	30	.3	tt—
*Bathroom Sink	1,860	4	3,250	2	2.2	3.0	9	.3	386	3.3	—
*Laundry Machine	7,900	16	20,300	17	35.3	48.5	316	9.8	4,790	40.4	—
Water Closet	23,540	48	67,780	57	16.0	22.0	2,782	86.5	6,473	54.5	—
Total Pollution	48,690	100	119,410	100	72.7	100.0	3,224	100.0	11,862	100.0	—

NOTE: Since no samples were tested from the water closet wastewater, the estimated human waste production values were used.

*Values were calculated from Table 3 using Table 2 weighted average water consumption rates.

[tt]Families A, B, and E were analyzed and the range of values of T.K.N. were 54 to 410 mg/c/d.

Table 9. Percent of Total Water Use for Each Plumbing Fixture

	Range	Mean	Ref. 7	Ref. 8	Ref. 6	Ref. 10
Kitchen	5 to 16	9	11	25	10	20
Bathroom	12 to 40	26	37	30	20	30
Laundry	4 to 27	18	4	5	25	20
Toilet	41 to 65	47	41	40*	45	30
Other	—	—	7	—	—	—

*Adjusted from 1.4 liters/flush to 14 liters/flush.

CHARACTERIZATION OF TYPICAL HOUSEHOLD GREY WATER

Warren D. Hypes

INTRODUCTION

The characterization of household grey water can produce as many profiles as there are family units generating it. At the least, there would be subtle differences due to different family living habits and the fact it is unlikely any two families use identical household and personal hygiene aids. Also, there is an unlimited variety of specialized cleansing agents, solvents, and other household related liquids whose presence would alter the chemical makeup of grey water. They would, of course, also alter the microbiological character of the water.

For the above reasons, the experimental program from which the following data were obtained attempted to characterize "typical grey water." Characterization of grey water was the first step in a recent experimental program conducted by the National Aeronautics and Space Administration's Langley Research Center, Hampton, Virginia, 23665. The interest was exploration of possible applications of spacecraft technology to domestic water and waste processing. In this program, characterization of grey water was accomplished to serve as a problem definition and to provide a baseline from which reclamation processes could be judged. Characterization, however, proved to be a valuable independent technology end point, and data related to characterization have been extracted and condensed for the following discussion.

Characterization of typical grey water needs as a goal, one necessary breakdown classification. Grey water can be characterized without garbage disposal solids and with garbage disposal solids. This breakdown is justified on the basis that many homes typically considered utilize this appliance and many do not. Inclusion or exclusion of the garbage solids has a large effect on the characterization.

GREY WATER WITHOUT GARBAGE DISPOSAL SOLIDS

This classification of grey water includes the discharge from the clothes washing machine, dishwasher, kitchen sink, tub/shower, and bathroom lavatory. During the experimental efforts, a typical makeup of household grey water was prepared each day for loading the processing systems. The daily makeup is shown on Table 1 and was centered around a family of four, two adults and

---

Warren D. Hypes, Aerospace Technologist, National Aeronautics and Space Administration, Langley Research Center, Hampton, Virginia 23665.

two children. The composition of the laundry and soiled kitchen utensils varied at random as the normal family living pattern dictated. Slight variations in the normal pattern were inserted by adding three additional shaves to the composition. This was done to get additional amounts of shaving creams into the waste waters, thus providing a more typical chemical characterization. The composition of all the cleansing agents was controlled to give typical data. Note that the cleansing agents used in the washing machine, dishwasher, and kitchen sink were composites. The use of the composites had the effect of averaging the variations in chemical compositions between different brands of cleansing agents.

During three tests of experimental waste water processing subsystems, nine separate accumulations of grey water with the makeup shown in Table 1 were generated. They were analyzed for the chemical/physical characteristics shown in Table 2. Also given on the table for comparative purposes is an average of three tap water samples taken from the experimental setup prior to initiation of the tests. As can be observed from the table, two of the ionic characteristics (phosphates and sulfate), all of the organic characteristics, and all of the physical characteristics (except pH) showed significant increases above the tap water baseline values. Several other characteristics including copper, lead, zinc, ammonia, chloride, and nitrate/nitrite showed some rise in characteristic values, but the slight rises are not considered significant. From the subjective viewpoint of acceptability, the most notable characteristics were turbidity, odor ("laundry"), and abundance of suds–reflected by the characteristic MBAS. The abundance of settleable particulates, to be discussed later, was also an obvious characteristic.

Samples for microbiological characterization of the grey water without garbage disposal solids, were also taken during the tests of the wastewater processing subsystems. Because microbial counts are so closely related to conditions prior to and during a specific test, and since a comparison between grey water with and without garbage solids was desired, the counts obtained during a later test that included both of these types of grey water were used for characterization, and they are discussed in the next section of this report.

GREY WATER WITH GARBAGE SOLIDS

This classification of grey water has the same composition as the previous grey water plus the addition of solids from a garbage disposal unit. Eight separate collections of grey water with garbage disposal solids were generated and analyzed. This grey water had the same constituents as before and was generated by the same procedures previously discussed. Fresh kitchen garbage scraps were accumulated daily with random composition. These compositions and the related volumes of water used to grind and transport them are shown in Table 3. The ground garbage and transport water were collected in a container before being dumped into the collected grey water. Samples for chemical/physical analyses were withdrawn immediately and were analyzed for characteristics shown in Table 4. As can be observed from Table 4, the average values of the organic and physical characteristics were very high with significant spread between low and high values.

After sampling, ground garbage and transport waters were discharged into and mixed with the other collected grey water, thus forming a collection of typical grey water with garbage disposal solids. Samples were then taken for chemical/physical and microbiological analyses. Results of these analyses are shown in Table 5. There it can be seen that several characteristics show significant increases with the addition of the garbage to the grey water. Increasing characteristics were BOD, COD, carbon chloroform extract (CCE–greases), and TOC. Increasing physical characteristics were conductivity, suspended solids, total solids, and turbidity. Two characteristics, phosphates and MBAS, showed a significant decrease with the addition of the garbage. Undoubtedly, the organic garbage particles absorbed or adsorbed the phosphates and MBAS and carried them down during settling. An additional sample taken after a 24-hour settling period showed decreases of 22% in BOD, 42% in COD, 29% in CCE (greases), 31% in TOC and 33% in suspended solids.

During generation and collection of the eight batches of grey water to which garbage was added, six of them were sampled for microbiological characterization. Prior to the test, the appliances and setup were rinsed with tap water. They were then filled with tap water chlorinated to approximately 5 mg/1 of chlorine and left overnight. The following morning, all the appliances and the setup were drained and then refilled with tap water. Samples were then taken from the collection tank for microbiological analyses to serve as a microbial baseline. As can be observed in Table 6, the assembly was sterile at the beginning of the test period. Immediately after generation of the grey water but prior to the addition of garbage, a sample was taken and analyzed for total organisms and coliform organisms. The low, high, and average of the samples from the six batches of grey water are shown in Table 6. Another sample was taken from each batch immediately after addition of the garbage and again after a 24-hour settling period. The six batches that were sampled correspond to garbage compositions 3-8 in Table 3.

The data in Table 6 show that counts of total and coliform organisms did not change significantly between the grey water with and without garbage samples. However, after a 24-hour hold in the collection and settling tank, both counts increased very significantly. The counts of 2.17×10^7 total organisms/ml and 5.40×10^8 coliform colonies/100 ml are the highest counts recorded throughout the test program.

SOME NOTES ON PARTICULATES

One characteristic that must be considered in any concept for disposal or processing of grey water is the content of undissolved particulates. None of the chemical/physical characteristic analyses used adequately quantified the magnitude of the particulate problem. The particulates in grey water without garbage solids originate primarily from the laundry, although some originate from bathing. Regardless of the type of clothing washed, an accumulation of lint and grit would appear on the bottom of the collection tank. In the 30-inch diameter tank used, the particles from one load of laundry and two baths would settle and accumulate into a layer from 1/8 to 1/4-inch thick. No effort to measure time of settling was attempted, but visual observation disclosed that it continued overnight. In addition, any agitation of settled matter would resuspend the particles. Most of the accumulation in grey water without garbage appeared to the eye as a light grey-colored lint. Under a microscope, the particles appear as a composite of fibrous lint, grit, and hair. During later tests of some processing systems, a quantification of the particulates disclosed an average of 44 grams dry weight of particulates per 100 gallons of mixed laundry and bath waters. The particles were also measured and categorized according to their maximum dimensions. The particle size distribution was as follows.

1 - 2 microns – 13 percent
3 - 10 microns – 75 percent
11 - 19 microns – 10 percent
Over 19 microns – 2 percent

Most of the particles settle to the bottom of the collection vessel during an overnight hold. If separation of the particles from the body of the grey water is desired, it can be relatively easily accomplished after a settling period, by drawing off the main body of the water leaving approximately two to five gallons (depending on tank bottom configuration) of water containing the particles in the bottom of the tank. This small volume of water can be dumped, and the drain action will carry the solids out of the tank. A shallow "V" bottom tank configuration aids the wash-out action.

Grey water with garbage has particulate characteristics similar to grey water without garbage except that the addition of the garbage increases the undissolved solids content. It is perhaps erroneous to regard ground garbage as particulates. Depending on the type of garbage and possibly upon the brand of disposal unit used, the garbage emerges in a range of consistency from a foam-like mush to hard, solid pieces. The majority of the ground garbage will settle to the bottom of a quiescent tank; however, a larger percentage of garbage particles than bath and laundry particles will float. The garbage particles also have a greater tendency to adhere to tank walls when the water is pumped or drained out of the tank. The tendency to adhere appears to be directly related to the amount of oily and greasy materials present.

Table 1. Daily Makeup of Typical Household Grey Water Without Garbage Disposal Solids

Appliance	Load	Cleansing Agent
Washing Machine	1 - Load Mixed Domestic Laundry	225 ml of mix 16[1]
Dishwasher	1 - Load Mixed Soiled Dishes	20 ml of mix 4 solid[2]
Kitchen Sink	1 - Load Mixed Soiled Pots and Pans	9 ml (1 tbsp.) of mix-4 liquid[3]
Tub/Shower	4 - Showers	1 - Handsoap A 1 - Handsoap B 1 - Handsoap C 1 - Handsoap D
	4 - Shampoos	1 - Shampoo A 1 - Shampoo B 1 - Shampoo C 1 - Shampoo D
Lavatory	4 - Shaves	1 - Cream A 1 - Cream B 1 - Cream C 1 - Cream D
	4 - Teeth Brushings	1 - Paste A 1 - Paste B 1 - Paste C 1 - Paste D

[1] Mix 16 - equal parts of 16 leading laundry detergents.
[2] Mix 4 solid - equal parts of four leading dishwasher solid detergents.
[3] Mix 4 liquid - equal parts of four leading dishwashing liquid detergents.

Table 2. Chemical/Physical Characteristics of Grey Water Without Garbage Disposal Solids

			Grey Water		
Characteristic	Unit	Tap Water	Low	High	Average
Arsenic	mg/1	<0.01	<0.01	<0.01	<0.01
Barium	mg/1	<1	<1	<1	<1
Cadmium	mg/1	<0.01	<0.01	0.03	0.01
Chromium	mg/1	<0.05	<0.05	<0.05	<0.05
Copper	mg/1	0.08	0.08	0.16	0.11
Iron	mg/1	0.18	<0.05	0.20	0.11
Lead	mg/1	<0.01	<0.01	0.10	0.04
Magnesium	mg/1	2.4	1.5	2.8	2.0
Manganese	mg/1	<0.05	<0.05	<0.05	<0.05
Nickel	mg/1	<0.05	<0.05	<0.05	<0.05
Selenium	mg/1	<0.01	<0.01	<0.01	<0.01
Silver	mg/1	<0.05	<0.05	<0.05	<0.05
Sodium	mg/1	8	68	93	80
Zinc	mg/1	0.39	0.37	1.60	0.62
Ammonia	mg/1	0.06	<0.05	0.80	0.18
Calcium	mg/1	24	15	17	16
Chloride	mg/1	19	20	30	25
Cyanide	mg/1	0.02	0.02	0.02	<0.02
Fluoride	mg/1	0.75	0.70	0.95	0.81
Nitrate/Nitrite	mg/1	0.2	<0.1	2.1	0.9
Phosphates	mg/1	1	50	68	59
Sulfate	mg/1	40	83	160	117
BOD	mg/1	*	270	360	328
CCE	mg/1	<10	11	41	20
COD	mg/1	12	283	549	452
MBAS	mg/1	<1	16	39	22
TOC	mg/1	<5	60	92	80
Color	$PtCl_6$ equiv. units	<5	30	>100	68
Conductivity	μ/mhos/cm	207	320	390	358
Odor	Threshold number	1	2	4	3
pH	pH units	7.2	6.9	7.5	7.2
Suspended Solids	mg/1	<10	17	68	33
Total Solids	mg/1	108	113	451	382
Turbidity	mg/ℓ, SiO_2 equiv.	1	30	68	49

*Not applicable

Table 3. Composition of Daily Garbage

Composition Number	Water Volume	Garbage[1]
1	2 Gallons	1 carrot, 1 banana peel, 3 lettuce leaves, 1 cup spoonbread, 1/2 grapefruit hull, 1 cookie, 1 piece bread, 1 tablespoon gravy
2	2 Gallons	Cucumber peel, 1 eggshell, onion peel, carrot scraps, 4 small lettuce leaves, 3-inch piece celery, 1 piece raw chicken skin, piece tomato, 3 pieces radish, 1 small slice ham–lean, 1/3 cup milk, 1 tablespoon ham fat
3	2½ Gallons	2 slices lime, tomato sauce, 2 pieces broccoli, 1 piece pork, 2 slices bread, 1/4 corned beef sandwich, 2 pieces chicken skin, 1/2 slice bread, 1/3 cup milk gravy, 1 tablespoon potato salad
4	2 Gallons	3 pork patties, 1/2 cup milk gravy, 1/2 teacup green beans
5	2½ Gallons	1/2 cup applesauce, 1 cup noodles, 1½ cup cereal milk, 1 slice bread, 2 lettuce leaves, 6 tomato slices
6	2 Gallons	3 slices bread, 1/2 sweet potato, 1 teaspoon cheese, 2 eggshells, 1/2 cup ham fat, 6 apple peelings, 1/3 cup chili, 2 cooked carrots, 1/2 biscuit
7	1½ Gallons	2 eggshells, 1 cup turkey scraps and skin, 2 potato peelings, 3 tomato tops (slices), 2 pieces onion skin
8	1¾ Gallons	6 slices tomatoes, 2 hot dog buns, 1/3 cup green beans, piece turkey skin, 1/3 cup turkey meat, several pieces onion skins, 1/2 cup chili

[1]The volume of unground garbage in all compositions, except number 4, was one quart. The volume of composition number 4 was one pint.

Table 4. Chemical/Physical Characteristics of Random Garbage Disposal Effluents

		Garbage and Transport Water		
Characteristic	Unit	Low	High	Average[1]
BOD	mg/1	3,000	8,100	5,114
CCE	mg/1	170	2,060	1,059
COD	mg/1	4,510	13,700	9,292
Conductivity	μ mhos/cm	540	9,400	2,134
Suspended Solids	mg/1	1,658	5,160	3,570
Total Solids	mg/1	3,162	10,171	6,604
Turbidity	mg/1, SiO_2 equiv.	160	1,500	614

[1] Average of eight samples

Table 5. Chemical/Physical Characteristics of Grey Water With Garbage Disposal Solids

	Characteristic	Unit	Tap Water	Grey Water Low	Grey Water High	Grey Water Average
Metals	Arsenic	mg/1	<0.01	<0.01	<0.01	<0.01
	Barium	mg/1	<1	<1	<1	<1
	Cadmium	mg/1	<0.01	<0.01	<0.01	<0.01
	Chromium	mg/1	<0.05	<0.05	<0.05	<0.05
	Copper	mg/1	0.08	0.14	0.20	0.17
	Iron	mg/1	0.18	0.19	0.64	0.46
	Lead	mg/1	<0.01	0.02	0.04	0.03
	Magnesium	mg/1				
	Manganese	mg/1	<0.05	<0.05	<0.05	<0.05
	Nickel	mg/1	<0.05	<0.05	<0.05	<0.05
	Selenium	mg/1	<0.01	<0.01	<0.01	<0.01
	Silver	mg/1	<0.05	<0.05	<0.05	<0.05
	Sodium	mg/1	8	71	78	75
	Zinc	mg/1	0.39	0.35	0.53	0.45
Ions	Ammonia	mg/1	0.06	0.05	1.50	0.49
	Calcium	mg/1	24	14	17	15
	Chloride	mg/1	19	25	45	33
	Cyanide	mg/1	<0.02	<0.02	<0.02	<0.02
	Fluoride	mg/1	0.75	0.73	1.00	0.87
	Nitrate/Nitrite	mg/1	0.2	<0.01	<0.01	<0.01
	Phosphates	mg/1	1	7	9	8
	Sulfate	mg/1	40	96	110	103
Organics	BOD	mg/1	*	210	650	480
	CCE	mg/1	<10	13	63	42
	COD	mg/1	12	276	836	582
	MPAS	mg/1	<1	1	9	4
	TOC	mg/1	<5	77	360	169
Physical	Color	$PtCl_6$ equiv. units	<5	>100	>100	>100
	Conductivity	μ/mhos/cm	207	340	460	417
	Odor	Threshold number	1	2	4	3
	pH	pH units	7.2	6.9	8.5	7.8
	Suspended Solids	mg/1	<10	72	183	115
	Total Solids	mg/1	108	435	654	559
	Turbidity	mg/1, SiO_2 equiv.	1	60	150	114

*Not applicable

Table 6. Microbiological Characteristics of Grey Water With and Without Garbage Disposal Solids

	Characteristic	Unit	Microbiological Counts in Grey Water			
			Tap Water	Low	High	Average[1]
Without Garbage	Total Organisms	cells/ml	0	1.0×10^5	4.4×10^6	1.50×10^6
	Coliform Organisms	cells/100 ml	0	$<1.0 \times 10^5$	7.7×10^7	1.95×10^7
With Garbage	Total Organisms	cells/ml	–	1.5×10^5	4.5×10^6	1.37×10^6
	Coliform Organisms	cells/100 ml	–	$<1.0 \times 10^5$	7.4×10^7	1.88×10^7
After 24 hrs. with Garbage	Total Organisms	cells/ml	–	2.3×10^6	5.4×10^7	2.17×10^7
	Coliform Organisms	cells/100 ml	–	$<1.0 \times 10^6$	1.3×10^6	5.40×10^8

[1] Average of six samples

SOIL MANTLE AS A PURIFICATION SYSTEM FOR GREY WATER

K.Y. Chen

INTRODUCTION

Soil as a treatment system has traditionally been exemplified by the extensive use of the individual septic tank with its appurtenant subsurface disposal system. The septic tank as used by the individual family contains comingled body wastes (also termed "black waste") and grey water, the latter that from showers, tubs, sinks, and other water-using household appliances and fixtures. Also, septic tanks are sometimes used for industrial wastes, even where centralized sewerage systems exist. When black wastes are separated from grey water, valuable drinking water otherwise used for flushing and transportation of waste can be saved. This separation also substantially decreases the loading of pathogenic disease-causing bacteria and viruses into the wastewater treatment and soil systems.

The functions of a grey water treatment tank are threefold: (1) the removal of solids, (2) storage and long-term digestion of sludge solids, and (3) separation of fluids and solids, including both floatable and settleable solids. The effluent fluid, which is relatively free of solids in comparison with the untreated household grey water, can be more readily infiltrated into soils. Despite this treatment, the effluent still contains relatively large quantities of impurities which will be subsequently removed in the soil system. These impurities could include infectious agents, nutrients, detergents, solid residues, and toxicants such as hydrogen sulfide and trace metals. The following section will discuss the removal and fates of such impurities from wastewater by the soil system.

PHYSICAL PROPERTIES OF SOIL SYSTEMS

Generally, soils are considered to be rather simple systems, while in fact, they are very complex in both physical and chemical makeup. Moreover, surface layers of soil are teeming with organisms. Therefore, to understand the soil mantle as a purification system, it is necessary to consider all of three properties of soils–physical, chemical, and biological.

The physical framework of soil consists of intermingled clay, silt, sand, gravel, stones, and organic matters. The discrete particles vary widely in size

---

Kenneth Y. Chen, Ph.D., Associate Professor of Environmental Engineering Programs, School of Engineering, University of Southern California, University Park, Los Angeles, California 90007.

and shape. The finest particles are clay at five microns or less in diameter (there are different schemes of classification) representing the main part of the colloidal materials of soils. Along with clay, organic matters, particularly humus, affect many of the properties of soil, especially the surface phenomena.

The particulate matter of soils constitutes texture, formal classifications of which depend on relative concentrations of different particle sizes. The arrangement of the particles into aggregates results in what is termed soil structure. Structure of a soil frequently determines its permeability, especially clayey soils.

Soils are very porous materials. About 40% of the volume of dry, compact sand and about 60% of dry, compact muck is air. By volume, about half of the most common soils–loams, silt loams, and fine sandy loams–consist of pore spaces of a cellular nature. This porosity is a most important quality affecting the movements of wastewater in soils and determining biological activities, the exchange of CO_2 and O_2, chemical reactions, and biogeochemical changes.

The principal movements of wastewater in soils are percolation, seepage, and movements due to capillarity and suction-pressure. Hydraulic gradients dictate preference to vertically downward flow, but less permeable substrata may cause lateral seepage. The rate of movement is rapid through coarse sand, and it is generally very slow through heavy silt loams and clays.

Soils are absorbent and adsorbent bodies, both physically and chemically. The principal sorptive components include colloidal clay and humus. The most important physical properties of soil systems for the removal of pollutants from grey water may be categorized as straining, sorption, flocculation, and sedimentation.

Straining takes place primarily at the surface of contact. Initially only those substances that are larger than the pore openings are removed. As straining continues, the strained substances will form smaller-sized pores, which in turn remove smaller particles from waste streams.

Physical sorption is one of the most important properties for the removal of both soluble and nonsoluble pollutants. This physical adhesion force is principally caused by the Van der Walls, hydrodynamic, and electrokinetic properties of the smaller soil particles.

The small pores of the soil provide opportunities for contact between the pollutants in wastes. As the pollutant floc builds up in size, it becomes large enough to be retained in the constrictions between individual pores.

Since the rate of flow in soil is usually very low, sedimentation may aid the removal of settleable particles in the soil pores. It is estimated that 400 sq.m. of effective surface area of a settling basin would be equivalent to a one-meter path through a fine sand.

CHEMICAL PROPERTIES OF SOIL SYSTEMS

Since coarse soil particles consist largely of unaltered or primary mineral particles, their chemical composition is similar to or the same as that of rock-forming minerals. Chemically, these coarse particles are almost inactive.

Clay particles differ from coarse particles in that they are products of chemical processes in rock weathering. They are the colloidal materials that comprise the active element of soil constituents.

Some clay compounds or minerals can be displaced by salts, acids, and bases without breaking down or changing their forms. This phenomenon, called ion-exchange, occurs in soils through interactions with prevailing anions and cations. The cation exchange process is recognized as the most important quality of soils. It is used to explain several important soil conditions, including soil acidity and alkalinity, the friability of some clays, fixation of potassium and ammonium, and non-fixation of nitrate, among others.

Besides chemical sorption and base-exchanging, many other ordinary chemical reactions take place in soils. For instance, some reducing soils may produce hydrogen sulfide. Through chemical reactions with hydrogen sulfide and precipitation, most trace metals in wastes may be deposited in soil pores. In that way, they are removed from percolating waste waters.

SOIL MICROBIOLOGY

If it were not for microorganisms, the earth's surface soon would be burdened with dead plants and animals in which the material bases of all life—carbon, nitrogen, phosphorus, and other elements—would irretrievably be locked up. In soil systems, these microorganisms live mainly on the surface of colloidal particles and partly in soil solutions. According to Waksman (1952) and Alexander (1965), soil organisms may be classified as given in Figure 1.

Investigators have found from 320,000 to 500,000 bacteria in a gram of sandy soil material, and from 360,000 to 600,000 in loamy material. In soils that are well supplied with organic matter, as many as 2,000,000 to 200,000,000 bacteria per gram of soil have been found. In humid regions, soil population is dense in the surface layers at depths of one to three inches. The number of microorganisms decreases in the deeper soil, almost disappearing at three to three and one-half feet. In the soil of arid regions, microorganisms may penetrate to a depth of six feet.

The effects wrought by soil bacteria, fungi, actinomycetae, algae, and protozoans are manyfold. Autotrophic bacteria, for example, utilize by-products formed by heterotrophic bacteria, including iron, manganese, ammonia and other nitrogen compounds, and hydrogen sulfide. In turn, heterotrophic bacteria utilize organic acids and other complex compounds formed by autotrophic bacteria. Thus, in a complex physical and chemical soil medium, microorganisms carry on their activities not as individual groups but together as a soil population. Moreover, a soil does not represent a homogeneous biological entity but rather a whole series of complex biological activities simultaneously in progress.

SOIL AS A TREATMENT SYSTEM

The soil mantle, with its immense volume and purification potential, is one of the most important natural treatment systems along with bodies of natural

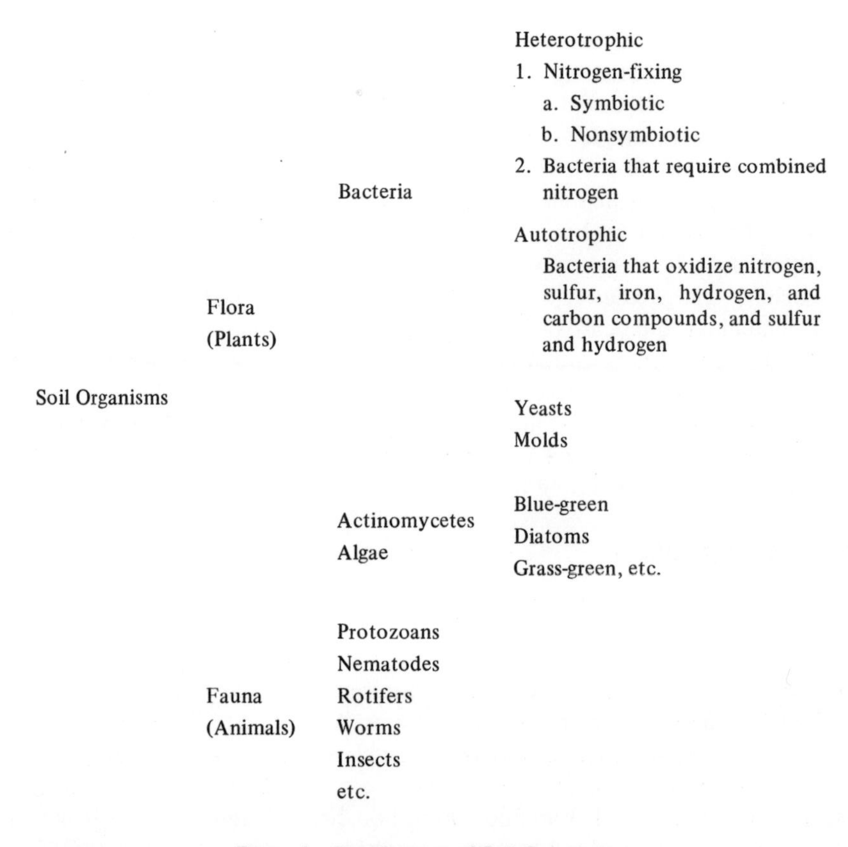

Figure 1. Classification of Soil Organisms

waters. Since ancient times man has used the soil mantle as a final discharge site for wastes. As previously mentioned, the soil mantle is characterized by many physical, chemical, and biological properties by which purification is made possible. By the straining, physical sorption, flocculation, and sedimentation mechanisms, most of the particulate forms of pollutants and some soluble contaminants can be removed. The chemical sorption, ion-exchange, and chemical deposition mechanisms can purify wastes of most soluble substances. Biological activities not only stabilize almost all the organic matters but also stabilize some inorganic substances.

However, the soil mantle also has its limitations in purification. Overloading or the inclusion of toxic substances in the wastes may cause failure in the purification mechanisms. Moreover, because of public health considerations, caution has been exercised in limiting waste discharge into the soil. Therefore, the suitability of the soil mantle as a treatment system is dependent on both the purification capacity of the soil and the characteristics of waste streams. The soil must have a safe treatment capacity, without interfering with ground water or impervious strata.

It was suggested in the *Manual of Septic-Tank Practice* (1967) that soil must meet two conditions, to be used as a conventional sewage treatment system:

A. The percolation time should be within the range specified in the following table:

Table 1. Traditional Septic Tank/Absorption-Area Requirements for Individual Residences (a) (Provides for Garbage Grinder and Automatic Clothes Washing Machines)

Percolation rate (time required for water to fall 1", in minutes)	Required absorption area, in ft^3 per bedroom (b), standard trench (c), seepage beds (c), and seepage pits (d)
1 or less	70
2	85
3	100
4	115
5	125
10	165
15	190
30 (e)	250
45 (e)	300
60 (e,f)	330

(a) It is desirable to provide sufficient land area for entirely new absorption systems if needed in future (i).

(b) In every case sufficient land area should be provided for the number of bedrooms (minimum of 2) that can be reasonably anticipated, including the unfinished space available for conversion as additional bedrooms.

(c) Absorption area is figured as trench-bottom area and includes a statistical allowance for vertical side wall area.

(d) Absorption area for seepage pits is figured as effective side wall area beneath the inlet.

(e) Unsuitable for seepage pits if over 30.

(f) Unsuitable for absorption systems if over 60.

(i) Sec. 5.1(b)(2)(A) p. 20 of Recommended State Legislation and Regulations: Urban Water Supply and Sewerage Systems Act and Regulations, Water Well Construction and Pump Installation Act and Regulations, Individual Sewerage Disposal Systems Act and Regulations, USPHS, HEW, July 1965.

B. The maximum season elevation of the ground water table should be at least four feet below the bottom of the trench or seepage pit. Rock formations or other impervious strata should be located at a depth four feet below the bottom of the trench or seepage pit.

The conditions just set forth are for domestic wastewater consisting of commingled black and grey water. For grey water alone, the strength free of toilet waste is reduced substantially with smaller volume. This feature presents the designer of a subsurface disposal system with a welcomed opportunity to cut the dimensions of the installation below USPHS recommendations for the same soil conditions or use a full-size treatment tank and soil disposal system to enhance system life. In any case, the general approach to using subsurface waste disposal techniques is as follows.

The percolation tests developed at the Robert A. Taft Sanitary Engineering Center help to determine general acceptability of the site and establish the design size of the subsurface disposal system. After making the percolation tests, the next information needed, but heretofore not readily available, is the quantity and quality of the sewage and the absorption area of the soil. When a soil absorption system is determined to be usable, three types of design may be considered: absorption trenches, seepage beds, and seepage pits (also disposal trenches, disposal beds, and disposal pits). How well any of these systems perform will be a function of the strength and quantity of the wastewater as we shall see in greater detail.

REMOVAL OF BIOLOGICAL AND CHEMICAL CONSTITUENTS

A. *Removal of Health Hazard Viruses and Bacteria*

Viruses and bacteria, pathogenic and otherwise, are quickly removed from wastewaters as they percolate through soils. Viruses are extremely small, electrically charged particles. Soil particles, especially clays, are also charged. Thus, a virus has not much chance of traveling more than a few inches through a soil than an iron filing would have of traveling through a gigantic pile of magnets. Once in soils, viruses have finite lifespans, and the soils thus offer a fantastically large capacity to assimilate them.

Bacteria are larger particles and are not as much affected by surface phenomena as are viruses. Thus, bacteria also are trapped by soils but might travel a few feet, depending on soil particle size, velocity of water movement, and other things. Still bacteria do not travel far and have a finite lifespan; so that, as is the case with viruses, soils have a very great capacity to assimilate bacteria.

The organisms most commonly used as indicators of the extent of water pollution from human and animal intestinal wastes are the coliform group. Orlob (1956) provided a formula by which the minimum distance between a septic tank and water supply may be estimated that will remove coliform bacteria and thus maintain drinking water standards. This equation is as follows:

$$L = 107\ dv$$

where L is the distance in feet, d is the effective diameter (D^{10}) of the soil in cm, and v is the ground water movement in ft/day. A conservative calculation with Orlob's equation leads to the conclusion that only a few feet are needed in typical soils to remove coliform bacteria adequately to protect drinking water quality. Since this distance is very small, the underground travel of bacteria should not be a concern. Soils are, indeed, very effective filters of bacteria.

B. *Detergents*

Synthetic detergents from home use can reach the grey water tank in slugs, which result in temporarily high detergent concentrations. On the average, the concentration of detergents in septic tanks is in the neighborhood of 25 mg/1, part of which is attached to sewage solids.

Winneberger and McGauhey (1965) showed that for water with a hardness of less than 100 mg/1, precipitation of detergents is generally small, regardless of excessively high detergent concentrations. As the hardness increases to about 200 mg/1, precipitation increases sharply at high concentrations of detergents.

At concentrations of surfactants generally encountered in household wastewater and at the most common water hardness levels, the threat of precipitated detergent adding to the clogging of soil systems is not serious.

C. *Phosphorus Removal*

Phosphorus removal from soil systems can be categorized into: (1) removal by microbiological means, and (2) removal by chemical or physical means (i.e., by chemical precipitation or sorption).

The effects of the microbiological removal of phosphate are significant only when artificial treatment systems, such as the activated sludge process, are applied (up to 90% phosphate removal), or when spray irrigation flow is used under aerobic conditions (76% to 93% phosphate removal) (Herbert, et al., 1965). It is interesting to note that the process of phosphorus uptake by microbiological means is in excess of the known stoichiometric ratio, which is not yet well understood. Chemical precipitation may account for most of the excess uptake.

The solubility equilibria and sorption phenomena were explained by Stumm in 1962, and Hsu in 1965. Many metal ions are particularly effective in precipitating phosphate from solution. The three metals contributing most heavily to phosphate precipitation are calcium, iron, and aluminum. All three of these metals commonly present in soils form relatively insoluble precipitates with phosphate.

Table 2 presents some chemical reactions for elimination of phosphates in sewage treatment plants (Stumm, 1962).

Table 2. Six Chemical Reactions Producing Relatively Insoluble Precipitates of Phosphate

$$Fe^{+3} + PO_4^{-3} = FePO_4$$

$$3Fe^{+2} + 2PO_4^{-3} = Fe_3(PO_4)_2$$

$$Al^{+3} + PO_4^{-3} = AlPO_4$$

$$Ca^{+2} + 2H_2PO_4^{-2} = Ca(H_2PO_4)_2$$

$$Ca^{+2} + HPO_4^{-2} = CaHPO_4$$

$$10Ca^{+2} + 6PO_4^{-3} + 2OH^{-1} = Ca_{10}(OH)_2(PO_4)_6$$

Soil systems chemically act similarly to the phosphorus removal reactions that occur in sewage treatment plants. The cations such as calcium, iron, and aluminum existing naturally in soil are particularly responsible for precipitating phosphate from septic tank effluent moving through soil (Webber, 1972). The efficiency of phosphate removal depends on the concentration of these cations in soil and on the pH conditions (Herbert, et al., 1965).

Hsu (1965) discussed the mechanism of phosphorus fixation in soils of different pH conditions. In acid soils, sorptions of phosphate on aluminum hydroxides and iron oxides, rather than solubility of phosphate compounds, seem to dominate. In neutral and basic soils, the iron phosphate and aluminum phosphate interactions are replaced by calcium phosphate. Solubility of metal phosphate probably dominates over sorption phenomena.

In a report on the use of the soil mantle as a wastewater treatment system, McGauhey and Krone (1967) state that "under normal conditions of soil pH, phosphates are effectively removed." Also, Weber (1972) came to similar conclusions on immobilization and fixation of phosphorus in soil attributing the immobility to the aluminum, iron, calcium, and organic compounds existing naturally in soil.

D. *Nitrogen Removal*

In grey water, the concentration of nitrogenous compounds is trivial in comparison with that of black water, an advantage of using a separate non-water using flushing toilet. Input from garbage grinders may account for the bulk of nitrogen in grey water.

The process of nitrogen fixation and denitrification in soil have long been familiar to plant physiologists. These processes are caused chiefly by the activities of microorganisms. Under either aerobic or anaerobic conditions, the organic nitrogen may be converted to ammonia by the action of saprophytic bacteria. The nitrosomonas group, known as the nitrite formers, can convert ammonia under aerobic conditions to nitrites:

$$NH_4^+ + \frac{3}{2}O_2 \xrightarrow{\text{Nitrosomonas}} NO_2^- + 2H^+ + H_2O$$

The nitrites are oxidized by the nitrobactor group of nitrifying bacteria:

$$NO_2^- + \frac{1}{2}O_2 \xrightarrow{\text{Nitrobactor}} NO_3^-$$

Under anaerobic conditions, nitrates and nitrites are both reduced by the denitrification process in the presence of electron donors (e.g., organic matters). Some nitrogen is thus reduced by bacteria to nitrogen gas, which escapes to the atmosphere, while a small fraction is reduced to ammonia.

Besides biological processes, the adsorption process is also an important factor in the removal of nitrogen. Preul (1968), studying the mechanism of the movement of nitrogen through soil, concluded that:

1. The main factors controlling the movement of nitrogen through the soil are adsorption and biological action.

2. When nitrogen is in the form of ammonium ions, physical adsorption on soil may be an important mechanism inhibiting the movement of nitrogen. Where nitrogen is in the form of nitrates, and within the normal

pH range of wastewaters, practically no inhibition to movement is offered.

In Preul's batch test, it was found that the adsorption of NH_3 by the soil followed the Freundlich isotherm equation and that a 90% equilibrium condition can be reached in about two hours. Under flow conditions through a soil bed, the rate of.adsorption for a soil may follow:

$$C/Co = e^{-(k/q)}$$

where C/Co is the ratio of NH_3 - N concentration in effluent and influent, k is a constant and q is the ratio of NH_3 - N in influent to the adsorbed amount.

Nitrogen usually exists in septic tank effluent in the form of organic nitrogen and ammonia; the usual reductions across the treatment device approximating 20% to 40% (Winneberger, 1973). Adsorption of nitrogen discharged to the soil mantle will occur and will be followed by biological effects. Thus, the total removal efficiency of nitrogen from septic tank/disposal field systems is about 40% to 72%, as estimated by Winneberger (1973).

Winneberger (1973) noted literature reporting high removals of nitrogen from sewage waters percolating through soil. Removals seemed to be more than expected, considering past engineering efforts to utilize denitrification. Winneberger postulated that denitrification in soils could result from microanaerobiosis; that is, nitrates could be carried by ground waters to a microanaerobic environment, such as a piece of decomposing carbonaeous root might offer. There, oxygen atoms of the nitrate molecule would be taken by bacteria for their needs. The nitrogen atom would then be either reduced to ammonia or combined with another nitrogen atom to form gaseous nitrogen in the soil atmosphere. Continued percolation of water would carry the ammonia out of the microanaerobic environment back into the aerobic environment, where oxidation to nitrate would occur. By percolation through the soil mass and the encountering of such microenvirons, the soils could offer a far more efficient denitrifying medium than would the simpler mechanisms engineers have devised.

In studies performed at Pennsylvania State University, chlorinated sewage effluent was applied by sprinkler irrigation at rates of one and two inches per week to areas of field crop and forest cover. Removals of 68% to 82% of nitrate nitrogen were reported.

At Santee, California, sewage effluents were placed in spreading basins, percolated through soils, and removed in a wastewater reclamation project. Removals of 30% to 40% of nitrogen occurred after 400 feet of underground percolation (Merrell, Jr., 1967), and total removal of nitrogen occurred after 1,500 feet of percolation (Bureau of Sanitary Engineering, California Department of Public Health, 1965).

Since grey water contains little, if any, nitrogenous materials, the traditional problem of an increase in nitrate level in ground water fed commingled sewage is greatly reduced.

E. *Other Organic Substances*

As discussed earlier, the activities of bacteria, fungi, algae, and protozoans can stabilize organic matters. When grey water is discharged to the treatment tank, the anaerobic bacteria convert some of the solid organic matter into liquids, which, upon discharge, diffuse into the soil for further stabilization.

The removal efficiency of organic matter by the soils is dependent on the characteristics of both the soils and the organic matter. Theoretically speaking, if the soil volume and degradation time are unlimited, almost 100% of the biodegradable organic matter can be removed. Therefore, removal efficiency will be limited by the purification capacity of the soil and the degree of biodegradation of the organic matter.

In general, the organic content of grey water is relatively lower in comparison with that of black water or other sewage in a sanitary sewer. To use the soil mantle as a treatment system for the removal of organic matter from grey water alone is not a serious problem. Only some relatively less degradable matters, such as detergents, oil, and grease in the grey water, may hinder removal; however, most of these substances are accumulated on the top part of a treatment tank and removed during periodic cleanings. Very little is discharged into the soil system under normal conditions.

F. *Removal of Trace Metals*

The surface adsorption of soil particles, especially clay minerals, can remove not only the colloidal form of trace metals but also the ionic form. The clay minerals constituting the main inorganic cation exchange materials include montmorillonite, illite, kaolin, and occasionally vermiculite or hydrobiolite (Bear, 1955). The cation exchange capacity ranges from one meq per gram of montmorillonite to less than 0.1 meq per gram of some kaolins. Comprehensive studies of the behavior of the trace metals in such exchange materials in soil are lacking. It has been shown that the order of exchange capacity is: clay > silt > sand. The metallic cations are very readily adsorbed by the exchange-active clay minerals and are correspondingly difficult to displace. The order of difficulty of displacement is approximately:

$$Cu > Pb > Ni > Co > Zn > Ba > Ca$$

Chemical precipitation followed by the straining effect or the adsorption effect can also remove some trace metals; e.g., trace metals can remove by ferric colloid, as has been described.

Since grey water contains extremely small amounts of trace metals, the removal capacity of the soil mantle is more than enough to reduce the pollution potential of trace metals.

SOIL CLOGGING

The percolation test is used to measure the rate at which water moves through soil. It can also be used to identify soils which might or might not be suitable for percolation systems. The conditions which govern the entrance of water into the soil surface are the controlling factors in percolation system design. It is assumed that small particles will pass through the matrix, whereas large particles removed by mechanical screening will clog its surface.

The factors contributing to soil clogging are usually classified as chemical, physical, and microbiological. But strictly speaking, clogging is a physical phenomenon resulting from the interaction or the integral of all three. Increased physical resistance to flow results from changed friction or viscosity coefficients or from reduction in size and volume of pore spaces. The practical consequences in the case of percolation systems derive from these and a number of other relationships. The biochemistry of aerobic versus anaerobic systems, hydraulic loading, organic loading, system geometry, and various operational procedures are among the most important aspects involved. Failure of a septic tank percolation system is, of course, a function of all factors in combination. However, it can be simply stated that the clogging phenomenon may be used as an indicator for judging the success of the treatment system.

Filtration of suspended solids is the first stage of soil clogging. Short-term soil clogging is sensitive to organic particles and inorganic ferrous sulfide suspended solids, whereas long-term clogging is more sensitive to operational procedures such as dosing and resting periods and the maintenance of the soil structure.

A. *Physical Factors*

Soils with a wide range of particle sizes may cause a higher degree of compaction than would soil of a more uniform particle size. Alteration of the surface structure by beating on the soil and the downward migration of fine particles to form a thin, compact layer on the soil surface will cause physical clogging. Orlob and Krone (1956) found that sedimentation of fine particles on the downstream side of larger particles, adsorption on the surface, and entrapment at the point of contact of larger particles leads to the clogging of soils even when clear water is applied. Clogging of this type, taking place over a period of time, becomes a permanent feature of the soil and governs its infiltration rate until some physical rearrangement of the soil mass occurs.

Another important physical factor in soil clogging is the retention of soil moisture by capillary forces. Winneberger, et al. (1961, 1962) showed that in a soil with grain size small enough surface tension and capillarity are major forces; there is a minimum length of soil column necessary to produce drainage once the application of liquid is stopped. This implies that water is essentially suspended in the soil by capillary forces. Oxygen may be excluded from the apparently unloaded field, and so an anaerobic condition may be maintained. This condition is not only favorable for anaerobic growths and ferrous sulfide formation (important clogging agents), but also decreases the rate of reopening of pore spaces by oxygen-using bacteria.

B. *Chemical Factors*

Some chemical reactions, especially ion-exchange, may produce percolation-system clogging. Best known is the deflocculation of soils by high concentrations of sodium. Fireman, et al. (1945), have shown that soils high in organic matter have higher permeabilities than do normal soils when irrigated with low-sodium percentage waters, and vice versa with waters of high sodium percentages. Further information about clogging effects caused by ion-exchange phenomena or other chemical effects is still lacking.

C. *Biological Factors*

Biological factors in soil clogging are affected by the properties of organic matter in wastes and the environmental conditions in soils. In general, biological factors in soil clogging only take place at or near the surface of contact. This surface clogging may increase by two methods:

1. Reduction of pore space by deposited organic suspended solids.
2. Reduction of pore space by bacteria growing on entrapped or dissolved solids.

The reopening of pore spaces may take place through bacterial decomposition of entrapped organic matter and decline of bacterial growth during drying or periods of inadequate substrate. A consideration of the relative rates of degreadation of organic solids aerobically and anaerobically suggests that aerobic conditions would take maximum advantage of the forces which tend to reopen clogged pores of the soil.

IMPROVEMENT OR REMOVAL OF SOIL CLOGGING

Improvement or removal of soil clogging to increase infiltration capacity is necessary for the success of grey water treatment. Most treatment failures are directly attributable to an inadequate knowledge of system design factors and soil requirements. Methods of preventing failures of the soil mantle purification system have been studied thoroughly by Winneberger and coworkers (1961-1965). The effective methods for the improvement of soil clogging may be categorized as follows:

A. Extra care in testing and constructing disposal fields to avoid destruction or puddling of the contact surface.

B. Resting the soil several days can do much to restore the original infiltration capacity of a sewage-clogged soil. In surface ponds, periodic drying can produce infiltration capacities above initial values as a result of an agglomeration of soil particles and subsequent changes in the structure of the soil surface. However, in a subsurface system, complete restoration of the infiltration capacity would require a longer period to overcome the anaerobic clogging. It is therefore suggested that the use of sectors of the system on a rotation basis is preferable.

C. In a subsurface system where drying is not feasible, soil drainage is necessary in order to restore aerobic conditions.

D. Effluents infiltrated laterally into a soil resulted in less clogging than that infiltrated upward or downward into the soil.

E. Experimental data show that flocculants can be used to lower suspended solids, COD, detergents, and other parameters of septic-tank effluent strength. However, the reductions had little or no beneficial effect on the soil-leaching system. Long-term field tests are needed to fully evaluate the usefulness of flocculants, particularly the possibility of beneficial effects on soil structures.

F. Use of a rock filter would take advantage of two phenomena: first, the tendency of bacteria and ferrous sulfide to form on the surfaces of stones immersed in wastes, particularly stones with rough surfaces; and second, the increased aggregation and sedimentation of suspended solids known to result from the shearing of water.

There is a continued need for knowledge of the biological, chemical, and physical processes of percolation field operations and the application of this knowledge to construction and design criteria. However, with sound information on the prevention of treatment failures, using the soil mantle as a purification system for grey water should be safe and feasible.

REFERENCES

Alexander, M., *Introduction to Soil Microbiology.* John Wiley & Sons, Inc., NY (1965, 3rd Ed.).

Anon., *Manual of Septic Tank Practice,* U.S. Department of Health, Education, and Welfare, Public Health Service No. 526 (1967).

Bear, F.E., *Chemistry of the Soil.* Reinhold Pub. Co., NY (1955).

Brandes, M., "Studies on Subsurface Movement of Effluent from Private Sewage Disposal Systems Using Radioactive and Dye Tracers." Ontario Ministry of the Environment, Interim Report, Part I (1972).

Bureau of Sanitary Engineering, State of California Department of Public Health, "Santee Filtration Study–A Study of Sewage Effluent Purification by Filtration through Natural Sands and Gravels of Sycamore Canyon at Santee." (1965).

Fireman, M. and O.C. Magistad, "Permeability of Five Western Soils as Affected by the Percentage of Sodium of the Irrigation Water." *Trans. Am. Geophysical Union, 26* (1945).

Herbert, B.F., Jr., P.C. Ward and A.A. Prueha, "Nutrient Removal by Effluent Spraying." *J. Sanitary Engineering,* Div. A.S.C.E., SA6, pp. 1-12 (Dec. 1965).

Hsu, R.H., "Fixation of Phosphate by Aluminum and Iron Acid Soils." *Soil Science, 99,* pp. 398-402 (1965).

McGauhey, P.H. and R.B. Krone, *Soil Mantle as a Wastewater Treatment System–Final Report.* Berkeley, U. of Calif., Ser. Rept. No. 67-11 (Dec. 1967).

McGauhey, P.H., G.T. Orlob, and J.H. Winneberger, *A Study of the Biological Aspects of Failure of Septic-Tank Percolation Fields–First Progress Report.* Berkeley, Sanit. Eng. Res. Lab., U. of Calif. (Dec. 1958).

McGauhey, P.H. and J.H. Winneberger, *Causes and Prevention of Failure of Septic-Tank Systems.* Tech. Studies Report, FHA, No. 533 (April, 1964).

McGauhey, P.H. and J.H. Winneberger, *Final Report on a Study of Methods of Preventing Failure of Septic-Tank Percolation Systems.* SERL Rept. No. 65-17. Berkeley, Sanit. Eng. Res. Lab., U. of Calif. (Oct. 1965).

Merrell, J.C., Jr., et al., *The Santee Recreation Project, Santee, California–Final Report.* Water Pollution Control Research Series Pub. No. WP-20-7, Cincinnati, Ohio (1967).

Orlob, G.T. and R.B. Krone, *Movement of Coliform Bacteria through Porous Media–Final Report.* Pub. Health Service Grant No. 4286, Berkeley, Sanit. Eng. Res. Lab., U. of Calif. (Nov. 1956).

Preul, H.C. and G.J. Schroepfer, "Travel of Nitrogen in Soils." *J. W.P.C.F., 40,* 1 (1968).

Stumm, W., "Chemical Elimination of Phosphates as a Third Stage Sewage Treatment." Discussion, International Conference on Water Pollution Research, Westminster, London (Sept. 1962).

Theis, T.L., P.C. Singer, W.F. Echelberger, and M.W. Tenney, "Phosphate Removal: Summary of Papers." *J. Sanit. Eng., 96,* SA4, pp. 1004-1009 (Aug. 1970).

Waksman, S.A., *Soil Microbiology.* John Wiley & Sons, Inc., NY (1952).

Webber, L.R., "Domestic Sewage Disposal in Soil." Dept. of Land Resources Science, U. of Guelph, Ontario (May, 1972).

Winneberger, J.H., "Nitrogen Contribution to the Truckee River from Tahoe Timber Trails." 1018 Hearst Ave., Berkeley, CA 94710 (Dec. 1973).

Winneberger, J.H., L. Francis, S.A. Klein and P.H. McGauhey, *Biological Aspects of Failure of Septic-Tank Percolation Systems–Final Report.* Berkeley, Sanit. Eng. Res. Lab., U. of Calif. (Aug. 1960).

Winneberger, J.H., P.H. McGauhey, *A Study of the Methods of Preventing Failure of Septic-Tank Percolation Fields–Fourth Annual Report.* Berkeley, Sanit. Eng. Res. Lab., U. of Calif. (Oct. 1965).

Winneberger, J.H., A.B. Menar, and P.H. McGauhey, *A Study of the Methods of Preventing Failure of Septic-Tank Percolation Fields–Second Annual Report.* Rept. of FHA, Berkeley, Sanit. Eng. REs. Lab., U. of Calif., (Dec. 1962).

Winneberger, J.H., A.B. Menar, and P.H. McGauhey, *A Study of Methods of Preventing Failure of Septic-Tank Percolation Fields–Third Annual Report.* SERL Rept. No. 63-9, Berkeley, Sanit. Eng. Res. Lab., U. of Calif. (Dec. 1963).

Winneberger, J.H., W.I. Soad, and P.H. McGauhey, *A Study of Methods of Preventing Failure of Septic Tank Percolation Fields–First Annual Report.* Rept. of FHA, Berkeley, Sanit. Eng. Res. Lab., U. of Calif. (Dec. 1961).

TD745 .M6
c.1
100107 000
Manual of grey water treatment
3 9310 00033163 5
GOSHEN COLLEGE-GOOD LIBRARY